The Shelduck
a study in behavioural ecology

The Shelduck
a study in
behavioural ecology

I. J. PATTERSON

ILLUSTRATIONS BY CHRIS FURSE

CAMBRIDGE UNIVERSITY PRESS
CAMBRIDGE
LONDON   NEW YORK   NEW ROCHELLE
MELBOURNE   SYDNEY

CAMBRIDGE UNIVERSITY PRESS
Cambridge, New York, Melbourne, Madrid, Cape Town, Singapore, São Paulo, Delhi

Cambridge University Press
The Edinburgh Building, Cambridge CB2 8RU, UK

Published in the United States of America by Cambridge University Press, New York

www.cambridge.org
Information on this title: www.cambridge.org/9780521113359

First published 1982
This digitally printed version 2009

A catalogue record for this publication is available from the British Library

Library of Congress Catalogue Card Number: 81–21231

ISBN 978-0-521-24646-0 hardback
ISBN 978-0-521-11335-9 paperback

Contents

Acknowledgments

A great many people have contributed to the work discussed in this book. I owe a particular debt to Maggie Makepeace, my research assistant for three years, for the unstinted effort and enthusiasm she gave to the Ythan shelduck study. Without her very capable fieldwork and analysis the results would have been very much poorer. Bill Murray gave willing assistance, especially with trapping and ringing the individually marked shelducks on which much of the Ythan study was based.

A number of postgraduate students at Culterty Field Station have made major contributions to the work on shelducks and have very kindly allowed me to quote from their PhD theses and other unpublished material. They include John Bateson, Nigel Buxton, Anat Gilboa, Stephen Redhead, Debbie Tozer, Murray Williams and Colin Young. Discussions with the staff and students at the Field Station, particularly during and after research seminars on the shelduck study, have greatly improved the interpretations and removed some of the worst errors. (The responsibility for the remainder, however, remains my own!) I have also had useful dialogues with shelduck workers elsewhere, particularly Peter Evans, John Hori, David Jenkins, Mike

Pienkowski and John Walmsley. Commander Chris Furse approached with enthusiasm his task of providing the drawings of shelduck, which contrasted somewhat with his more usual Antarctic subjects.

I am grateful to Professors G. M. Dunnet and W. Mordue of the Zoology Department of Aberdeen University, for facilities and encouragement; to the Nature Conservancy Council for permission to work on the Sands of Forvie and Ythan Estuary National Nature Reserve, and to the Director and staff of the Aberdeen Computing Centre for computing facilities. The Ythan shelduck study was supported for three years by grant GR3/2439 from the Natural Environment Research Council.

Finally, not the least of my debts is to my wife Muriel, who ungrudgingly tolerated the long unsocial hours I spent on shelducks and encouraged me during the writing of this book.

1

Introduction

Behavioural ecology, as its name suggests, has arisen as a hybrid field which incorporates the ecological and selective aspects of animal behaviour and the behavioural aspects of animal ecology. As such, it is concerned with the interrelationships between three sets of factors (figure 1.1): environmental ones such as food, nest sites or shelter, behavioural ones such as territorial aggression or dominance and population variables such as density, reproduction or mortality.

Earlier studies of population ecology emphasised the numerical relationships between variables (figure 1.1, A) such as the correlation between reproductive rate and subsequent changes in density and the effects of the environment on these. Lack (1964) for example showed that changes in the breeding density of great tits *Parus major* were correlated with the juvenile to adult ratio in the previous winter. This ratio in turn depended on the survival of juveniles in late summer, which was probably related to the production of insects. In much of such quantitative population ecology there was little or no consideration of the individual animal or of social behaviour.

Similarly a great deal of the study of behaviour has not been concerned with ecological or functional aspects of the subject, concen-

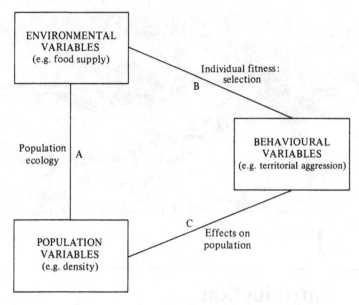

Figure 1.1 Relationships between fields of study in behavioural ecology. Re-drawn from Patterson (1980).

trating instead on such problems as causation, coordination, evolution and ontogeny of behaviour and its physiological and neurological basis (e.g. Hinde, 1970).

Behavioural ecology differs from its two parental disciplines by concentrating on the relationship between behaviour and the environment and that between behaviour and population variables (figure 1.1, B, C). It is crucial that these two subdivisions be clearly distinguished, since their theoretical bases are very different.

The study of behaviour and environment (figure 1.1, B) involves the idea of function; the survival value of the behaviour to the animal. Studies in this field commonly ask how an animal's behaviour is adapted to the environment so as to maximise the genetic fitness of the individual, or how an animal might adopt the best strategy for its optimal exploitation of some resource. For example Krebs (1978) showed that great tits select the size of prey taken so as to maximise their rate of food intake, and Pulliam (1980) has shown a similar result in seed selection by gramnivorous birds. Davies (1980) similarly argued that territorial aggression in several species is adapted to prevailing food density such that the occurrence of territorial behaviour and the size of the territory defended are both 'economic' in terms of return for the energy expended in defence. In all such studies there is a clear implication that selection has operated to adapt individual behaviour to the environment.

In contrast, studies of behaviour and population (figure 1.1, C) are concerned with the effects or consequences of behaviour on population variables, without implying that such effects have necessarily been selected for as the main function of the behaviour. For example, in the red grouse *Lagopus lagopus* year-classes were found to differ in aggressiveness, which was affected both by nutrition and genetically during cyclical changes in numbers. The different year-classes took different territory sizes in autumn, with consequent variation in breeding density (Watson, 1967; Moss & Watson, 1980). Similarly Murton (1968) showed that a consequence of individual striving for dominance over food in woodpigeons *Columba palumbus* was that flock size became adjusted to food abundance through the emigration of subordinate birds.

These two main approaches in behavioural ecology, adaptedness of behaviour and effects on populations, are closely linked in the sense that most behaviour affecting population variables can be explained in terms of individual selective value. In spite of Wynne-Edwards' (1962) major contribution in pointing out and emphasising the important effect of behaviour on populations, there has been little support for his secondary suggestion that selection operates at the population level. It can be argued, for example, that if territory size is adapted to food abundance so as to maximise the genetic fitness of the individual animals, then population density will vary with food supply as a necessary consequence, resulting in an adaptive change in density with change in resource level. Such explanations, based on selection for individual genetic fitness, have the advantage of being more easily explained in genetic terms than those based on population processes such as group selection (Wynne-Edwards, 1962).

The growth of theoretical discussion on behavioural ecology has emphasised the need for detailed field studies, to test some of the ideas on wild populations of animals in relatively undisturbed environments. The shelduck *Tadorna tadorna* is an excellent study species, being a large and conspicuous bird with complex and varied social behaviour but with small accessible local populations, allowing the relationships between behaviour and population to be explored fairly readily.

Most of the discussion in this book is based on material from a small number of detailed studies of the ecology and behaviour of the shelduck, principally those made on the Ythan estuary, Aberdeenshire, by myself and a number of colleagues since 1960 (Young, 1964a; Williams, 1973, 1974; Makepeace, 1973; Patterson, Young & Tompa, 1974; Buxton, 1975; Patterson, 1977; Patterson & Makepeace, 1979; Makepeace & Patterson, 1980). Other major studies are those of Hori (1964a, 1969 and others) on the Isle of Sheppey in the Thames estuary and those of Jenkins and his colleagues at Aberlady Bay on the Firth of Forth (Jenkins, 1972; Jenkins, Murray & Hall, 1975), later continued by Evans and Pienkowski (1982; Pienkowski & Evans, 1979, in press a). Early observations and counts on the Tay by Boase (1935, 1938, 1950, 1951, 1959, 1965) formed a useful basis for all the later studies. Unfortunately these study areas are not distributed evenly around Britain; all four are on the east coast, and three are in Scotland (see figure 2.9). However other studies of a number of particular aspects of shelduck ecology have been carried out on the west coast, e.g. observations on moult migration at Morecambe Bay (Coombes, 1949, 1950), the Cheshire Dee (Allen & Rutter, 1956, 1957, 1958) and the Severn (Morley, 1966), and counts of moulting flocks at Bridgwater Bay, Somerset (Perrett, 1951, 1953; Eltringham & Boyd, 1960, 1963; Eltringham, 1961).

This book will deal with a number of issues in behavioural ecology, particularly the effects of social behaviour on the limitation of population size, using the detailed studies of the shelduck to test and illustrate these ideas. My approach is slanted somewhat towards ecology rather than behaviour; for example, I have not attempted a detailed description and analysis of displays and postures, courtship or pair formation. After briefly introducing the bird and its general ecology, the main body of the book is divided by the main phases of the annual cycle of the shelduck, since its social organisation changes dramatically between seasons, raising different behavioural and ecological questions. Each chapter integrates population, behavioural and theoretical material relevant to that phase of the season, and discusses questions of both the adaptedness and population consequences of the social behaviour seen at that time. The concluding chapter attempts to bring together the preceding ones around the question of the limitation of population size and the role of social behaviour in this.

2

The shelduck

2.1 Taxonomy

Shelducks are intermediate between the true ducks and the geese, although they are usually placed with the ducks in the family Anatidae, where they join the sheldgeese in the tribe Tadornini (Johnsgard, 1961). There are seven species of shelducks: the common or northern shelduck *Tadorna tadorna* (L.), the ruddy shelduck *T. ferruginea* (Pallas), the Cape or South African shelduck *T. cana* (Gmelin), the Australian shelduck or mountain duck *T. tadornoides* (Jardine & Selby), the New Zealand or paradise shelduck *T. varietaga* (Gmelin), the radjah shelduck or burdekin duck *T. radjah* (Garnot) and the crested or Korean shelduck *T. cristata* (Kuroda) (which may be extinct). All are large brightly coloured ducks with many goose-like features. The lack of a camouflaged female plumage, the persistence of the pair bond and the prolonged parental behaviour shown by the male all resemble the geese, whereas the general morphology, voice and the existence of an eclipse plumage are duck-like. Johnsgard (1978) considered that, like the radjah shelduck, the common shelduck represents a rather isolated offshoot from the rest of the group. Both species are adapted for dabbling and feed chiefly on molluscs and other invertebrates, whereas the rest of the group are principally vegetarian.

The common shelduck was originally called *Anas tadorna* by Lin-
naeus in 1758, but has had a variety of other names including *Tadorna
cornuta* (Saunders, 1889) *Tadorna vulpanser* and *Tadorna bellonii*
(Yarrell, 1843). There seems general agreement that the common name
(and its older form sheldrake or shieldrake) was derived from 'sheld'
meaning pied, though Yarrell (1843) suggested 'shell' as an association
with molluscs in the diet, or even 'shield' from the use of the bird in
heraldry, as by the Brassey family in Hertfordshire! Other old names
include burrow duck, barrow duck or bar gander, presumably from the
habit of nesting in burrows, skeeling goose (Scotland) and sly goose
(Orkney). The shelduck is known as *Brandgans* in German, *tadorne de
belon* in French, *Bergeend* in Dutch and *tarro blanco* in Spanish.

2.2 Description
The shelduck is a large duck, around 60 cm long with a wing
span of about 120 cm. The male is noticeably larger than the female,
having a mean wing length of 334 mm (range 312–350) and mean
weights (in different areas) of 1167–1260 g, compared to the female's
wing of 303 mm (range 284–316) and mean weights of 813–1043 g
(Cramp & Simmons, 1977). The male appears taller and longer legged
than the female, particularly during the breeding season.

Both sexes have an unmistakable, boldly patterned plumage. The
head is green-black, usually with white on the forehead of the female,
which may also have variable pale areas or spots on the sides of the face
allowing different individuals to be distinguished. The neck is white,
ending in the characteristic bright chestnut breastband, which is
broadest and most intensely coloured in the male. The body is white
with black scapulars and a broad black belly stripe, again widest and
darkest in the male, where it is conspicuous in upright postures. The
under tail coverts are cinnamon, darker in the male than the female.
The wing coverts of adults are white, while the primaries and secondaries are black, with the latter strongly iridescent green on their outer
webs. A group of outer tertials with deep chestnut outer webs completes
the colourful effect of the plumage. The bill is bright red with a black
nail, although females may have a variable amount of slate grey at the
tip. The male has a large soft bill knob in the breeding season, the size
varying with testis size (Young, 1964a) showing it to be under hormonal
control. The legs and feet are pink or red and like the bill are more
deeply coloured in the breeding season.

Why should the shelduck be so conspicuously patterned, and moreover why should the sexes be almost identical, when most female ducks have inconspicuous plumages? The bright plumage may well be correlated with the shelduck's aggressive and territorial behaviour. Both sexes are highly aggressive and their threat displays are enhanced by the plumage pattern. The female takes an active part in territorial defence, and plumage similar to that of the male might be advantageous in disputes between pairs. The habit of nesting in burrows greatly reduces the disadvantage of having a conspicuous female on the nest and presumably evolved along with the plumage.

The annual moult for most shelducks starts in late June or early July, although those rearing young delay until late July or early August when the ducklings fledge. The process starts with a body moult into the eclipse plumage, which is not dark or dull as in most other ducks but is similar in general pattern to the normal plumage. The head becomes more brown with some white areas, and the bright chestnut breast feathers are replaced by brown ones with dull black subterminal bars and narrow white edges, giving a duller but still conspicuous plumage. The flight feathers are all moulted simultaneously during a flightless period of about a month between July and October, after the moult migration (see section 2.8). The normal breeding plumage is regained by a partial body moult before the birds return to the breeding areas.

The downy young have a bold 'double cross' pattern with a dark grey-brown stripe extending from forehead to tail with cross-stripes over the wings and pelvis, contrasting with the pale grey-white of the rest of the down (figure 2.1). The bill and legs are green-grey. As the down is gradually lost a black spot first appears on the cheek at three to four weeks of age, and then the duckling gradually assumes the juvenile plumage which is grey-brown on the head, back of the neck and upper parts, and white below, with some white around the base of the bill. There is usually no trace of the adult breastband or belly stripe, although some ducklings that I reared on commercial turkey chick food developed a diffuse brown band on the chest.

The juveniles have a body moult in September to October into a first winter plumage very similar to that of the adult. The wing coverts, however, are grey rather than white, and there are conspicuous white tips on the secondaries and inner primaries, giving a diagnostic white trailing edge on the wing, easily visible in flight (Hori, 1965a). The legs are usually grey with little pink, and the bill tends to be paler than the adult's.

The voice of the shelduck differs markedly between the sexes. The

Figure 2.1 Hand-reared shelducklings, with foster mother (Maggie Makepeace).

male has a variety of pleasant whistling notes, particularly a two-note whistle given in flight, a rhythmic series of short soft whistles given with agonistic displays and a long soft trill given with a ritualised shake, the Whistle-shake display. A number of other short whistles are difficult to distinguish from each other. The female has a much louder voice than the male, with a number of quacking calls. The commonest is a prolonged rhythmic call 'ak-ak-ak-ak-' (Witherby *et al.*, 1939), commonly used as a contact call, e.g. by a female leaving the nest to rejoin her mate. A lower growling or barking quack is used with the Inciting display during agonistic encounters, and a softer 'ga-aank' is used as an alarm call toward predators near the nest or young. A number of soft quacks and rhythmic calls are used by females while tending ducklings. The downy young have a chirring trill when in contact with each other or the parents and a loud piping, with a rhythm very similar to the long call of the female, when they become separated. The ducklings' voices 'break' to the adult form around or soon after fledging. More detailed descriptions and sonographs of the voice, together with body measurements and complete descriptions of plumages can be found in Cramp & Simmons (1977), Bauer & Glutz (1968) and Witherby *et al.* (1939).

2.3 Shelduck weight

Large numbers of shelducks were weighed when they were trapped to be marked on the Ythan estuary, Aberdeenshire, between January and June 1969–80. Changes in weight over the six-month period were measured by taking the mean weight of birds caught in each month. Only the first capture of each bird in each year was used to avoid any effects of repeated capture of the same individual (see below).

Seasonal changes in weight, January to June

Males showed a steady decline in weight from about 1350 g in January to about 1100–1150 g in May, with no further decline between May and June (figure 2.2). Males caught in May were significantly lighter than those first caught in January (territorial males, $d = 3.42$, $P < 0.001$; non-territorial males, $d = 6.43$, $P < 0.001$). Females did not change weight between January and May, although birds caught in March were somewhat lighter than those in February or April and there was some decline from April to June (figure 2.2).

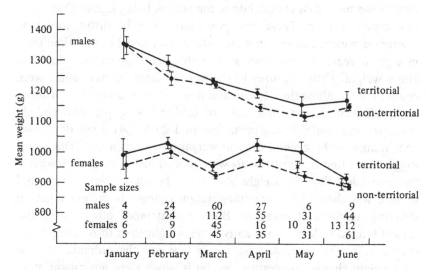

Figure 2.2 Changes in shelduck mean weight from January to June 1969–80 on the Ythan. Vertical bars are standard errors. Crosses show mean weight of females with brood patches.

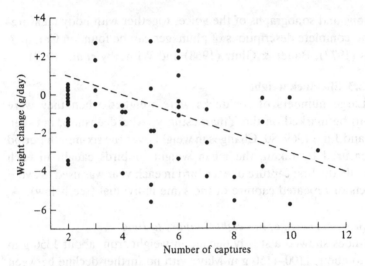

Figure 2.3 Changes in weight of individual shelducks in relation to the number of times they were captured during the period January to June. The dashed line is the calculated regression ($y = 0.162 - 0.322x$, $r = 0.364, p < 0.02$).

The decline in male weight between January and May, averaging 1.62 g per day in territorial males and 1.89 g per day in non-territorial males, could be due either to a seasonal loss of weight by each individual or to those birds first caught late in the season being lighter than those first caught earlier. These two possibilities can be distinguished by examining weight changes in males which were caught more than once in a given year, at least if we assume that repeated capture does not affect weight. Only captures which were at least 30 days apart were considered, to allow time for measurable weight change; 19 territorial males showed a mean loss rate of 0.91 ± 0.45 g per day and 32 non-territorial males had a mean loss of 1.29 ± 0.41 g per day, rates comparable to the general drop in weight of new captures. This result suggested at first that individuals were indeed losing weight. However, the possibility that the weight loss might be an artifact of repeated capture was checked by examining weight changes in males caught on differing numbers of occasion. Birds caught more often during the season tended to lose significantly more weight than those caught less (figure 2.3). The calculated regression line cuts the ordinate close to zero weight change, suggesting that birds which were not caught at all might not lose weight. (However, the relationship for the trapped birds may not hold for those never caught.) A change in weight following

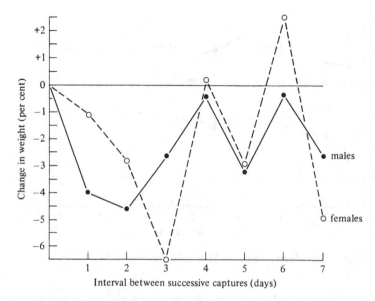

Figure 2.4 Changes in mean weight in relation to the interval between successive captures.

trapping was confirmed by examining weight changes in birds recaptured a varying number of days after their first capture. Males lost a mean of 4.0 per cent of their weight by the day after their first capture, reached a peak of 4.6 per cent after two days but had returned to around their original weight by four days later (figure 2.4). Females showed a similar though more severe loss, of 6.5 per cent of their original weight after three days. Both sexes showed puzzling parallel fluctuations in weight in those birds caught four to seven days after their first capture (figure 2.4).

It seems clear from these results that the weight loss in recaptured Ythan shelducks was related to the trauma of capture, or that the birds which were losing weight were caught more often. There was no remaining evidence of seasonal weight loss by individuals, so that the observed decline in monthly mean weights must have been due largely to differences in the weight of birds being captured for the first time in different months. This is more probably due to heavier categories of bird arriving earlier in the breeding area.

Weights of different age and social categories
The captured Ythan shelducks could usually be assigned only a minimum age, since most had first been marked as adults, at least two

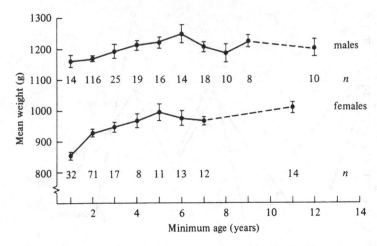

Figure 2.5 Mean weights of Ythan shelducks of different ages.

years old. Both sexes showed some increase in weight with age (figure 2.5), males reaching peak at six years of age with no further increase and perhaps a slight decrease in older birds. Females reached a similar peak at five years, although old females, over eight, were slightly (but not significantly) heavier than those of six or seven (figure 2.5). Females known to be one year old were significantly lighter than those of at least two ($d = 3.58, P < 0.001$). Since older shelducks usually arrived on the Ythan earlier than younger ones (Williams, 1973, and figure 3.2) the greater weight of older birds can at least partly explain the seasonal decline in the weight of newly captured shelducks.

In every month non-territorial shelducks of both sexes were lighter than those which were territorial in the year they were weighed (figure 2.2). The differences were mostly small and were statistically significant for males only in April ($d = 2.34, P < 0.02$) and for females only in May ($d = 2.12, P = 0.05$). Young (1970a) also found that territorial males were heavier than non-territorial males. He attributed the difference mainly to the greater age of territorial males (although he set out to test the possible advantage of being territorial). However, in the 1969–80 data, even within males of the same age, there was a tendency for the territorial ones to be heavier. In each year class, from two to five years of age, territorial males were heavier than non-territorial ones (although none of the differences was significant). Whatever the cause of the greater weight of territorial shelducks, it is unlikely to be a consequence of having a territory. The difference was already present in January to

March before the territories were occupied (figure 2.2). There was a greater difference after territory establishment in April, especially in females, but this may well have been due to weight gain associated with breeding, since non-territorial birds do not breed. It seems likely that it is the fittest and heaviest birds which succeed in the competition for territories. Coulson (1968) similarly found that kittiwakes *Rissa tridactyla* in the central parts of a colony were larger and heavier than those nesting on the periphery and that the difference was apparent when the birds first arrived. Since more non-territorial than territorial birds were caught in May and June (sample sizes in figure 2.2), the weight difference between categories will also contribute to an overall seasonal decline in the weight of newly caught shelducks.

Thus the initial picture of loss in male weight with season seems to be explained by changes in the categories of bird caught in different months, with older, territorial birds being both heavier and earlier in arriving. There seems no need to look for adaptive change in the weights of males through the winter. However, it may also be advantageous for shelducks to carry larger fat reserves in the coldest months with the greatest risk of periods of starvation, and it may be only the heavier males which can risk moving to small northern breeding areas early in the winter.

Finally, some incubating females, with functional brood patches, were caught in May and June when they visited traps between incubation bouts. These birds were intermediate in weight between territorial and non-territorial females (figure 2.2), suggesting a small, but not statistically significant, weight loss when incubating. The captured birds could have gained from the high-energy food supplied at the trap but this would be offset by the trauma of capture discussed earlier.

Weight during the moult migration

Cramp & Simmons (1977) presented data from the Netherlands, Denmark and Kazakhstan, USSR, on shelduck weights in June to August. The mean weights were 1167 ± 129 g ($n = 7$, range = 1000–1350) for males and 952 ± 88 g ($n = 5$, range = 850–1075) for females, and so were similar to those measured on the Ythan in June (figure 2.2). Dementiev & Gladkov (1952) similarly listed weights of 1000–1150 g for males in summer and 1000 g for females in June in the USSR.

Oelke (1969*b*) weighed moulting shelducks on the Grosser Knechtsand and found that 11 adult males just starting to moult weighed 1430 ± 170 g (range = 1180–1710) but that those further into the moult were significantly lighter. Thus 13 males with new feathers at least half-grown weighed only 970 ± 190 g (range = 750–1475). Similarly, 14 adult females at the start of moult weighed 1105 ± 110 g (range = 860–1250) whereas eight with new feathers were significantly lighter, at only 930 ± 180 g (range = 690–1180). Yearlings showed similar decreases in weight during the moult and were lighter than adults at the same stage (significantly so in males).

These data suggest that shelducks lose condition while moulting (unless different categories of birds were caught at different stages), presumably because the energy required to grow new feathers out-weighs the net intake from feeding. Heat loss may also be increased through the moulting of body feathers. After the moult the birds must regain their reserves before returning to their breeding areas, since the earliest arrivals at the Ythan were heavy (figure 2.2).

2.4 Diet

The shelduck feeds almost entirely on small invertebrates, chiefly molluscs, crustaceans and insects, often on species which are surprisingly tiny for such a large duck. The diet varies considerably across the range, with a preponderance of molluscs taken by the western birds and crustaceans and insects by the eastern ones.

In Britain, the commonest food item is the tiny estuarine snail *Hydrobia ulvae*. Olney (1965) found it in all 46 birds he examined at a number of English sites throughout the year. In 18 birds, from the Medway in Kent, *Hydrobia* made up 89.5 per cent of the volume of food, and one bird contained over 3000 snails. There were small quantities of the small amphipod crustacean *Corophium volutator* (7.7 per cent of food volume), Jenkin's spire shell *Hydrobia jenkinsii*, the ragworm *Nereis diversicolor* and a small clam *Macoma baltica*. Plant material made up 7.3 per cent of the volume of the Medway sample but only 1.9 per cent of another of seven birds from Bridgwater Bay in Somerset, the commonest plant being the green alga *Enteromorpha* together with traces of seeds of marsh plants. Campbell (1947) found that one bird from North Uist in northwest Scotland had similarly eaten mainly *Hydrobia ulvae* with one rough periwinkle *Littorina saxatilis*. Buxton (1975) also found that on the Ythan estuary, Aberdeenshire, *Hydrobia ulvae* was the most important food, with some *Corophium* and *Nereis* (taken particularly by ducklings). One female returning to incubate after a one-hour feeding bout contained 11 858 individual

Hydrobia ulvae. Evans *et al.* (1979) considered small oligochaetes to be the most important dietary item at Teesmouth.

On the European moulting grounds at Heligoland Goethe (1961*b*) found that 12 birds in September and October again contained chiefly *Hydrobia ulvae* with some small *Macoma baltica* and sand gapers *Mya arenaria*. Droppings found on the mudflats or collected from captured birds contained numerous *Macoma baltica* and *Mucula catena* with some cockles *Cardium edule*, mussels *Mytilus edulis*, periwinkles *Littorina littorea* (once) and *Hydrobia ulvae* (once). The low incidence of *Hydrobia* in this sample was probably due to digestive breakdown, since all the molluscs were described as being broken into fine particles and the smaller, more delicate shells of *Hydrobia* are less likely to survive.

The food in the eastern part of the range, summarised by Dementiev & Gladkov (1952), contrasts with that in the west. An important food in several localities in summer was the brine shrimp *Artemia salina* and its eggs, with an average of 1800 shrimps (maximum 19 800) and 10 150 eggs (maximum 36 800) per shelduck gut examined. Midge larvae (Chironomidae) were also important, with an average of 5970 and a maximum of 68 880 per stomach! Single-celled algae, probably *Aphanoteca*, and swarming ants (Formicidae) were taken in some areas, and ducklings were found feeding on locusts *Calliptamus italicus*. Less appears to be known about winter food in the east, but vegetable material, including algae and seeds of water plants, e.g. *Ruppia*, seems to be more important there, although some molluscs (*Hydrobia* and *Theodoxus*) were taken (Dementiev & Gladkov, 1952). Throughout the range a number of other food species, listed by Cramp & Simmons (1977), have been recorded, although none appears to be of general importance.

2.5 Feeding behaviour

Buxton (1975) found that shelducks on the Ythan, Aberdeenshire, had a pronounced tidal rhythm of feeding by night as well as day, with most done on the flowing or ebbing tide and less at low and high water (except at neap tides). Bryant & Leng (1975) on the Forth near Edinburgh, although finding a similar peak of feeding on the flow tide, extending to high water, found that fewer birds fed on the ebb tide and low water. The birds presumably feed at tidal stages when *Hydrobia* is most easily available on or near the mud surface (Olney, 1965). Newell (1962) has shown that the snail's behaviour changes markedly with tidal state, most being on the surface or floating during the period of water

Figure 2.6 Feeding by sieving (male, on the right) and by pecking (female, on the left). Drawn from a photograph.

coverage. Shelduck usually feed in those parts of the mudflats where *Hydrobia* is most abundant (Bryant & Leng, 1975) and Buxton (1975) found that birds in areas with high densities of *Hydrobia* fed for less of their time than those in lower density areas. Shelduck at Teesmouth, where the principal food items were small oligochaetes, concentrated where these were most abundant (Evans *et al.*, 1979).

A number of different feeding methods are used depending on the prey and the water depth (Buxton, 1975; Bryant & Leng, 1975). The most characteristic shelduck feeding technique is sieving or scything (figure 2.6) where the front section of the beak is placed on the mud surface and swung from side to side as the bird walks slowly forwards. Mud and water can be seen gushing from the rear sides of the bill, presumably pumped by the tongue and small movements of the mandibles. Coarse lamellae at the tip of the bill possibly reject stones and larger items, while fine lamellae at the rear and on the edges of the tongue retain the invertebrates. This technique is used on exposed wet mud, while dabbling at the water's edge and presumably while head-dipping or upending in deeper water (Bryant & Leng, 1975). Fern-like trails may be left in the mud (Swennen & van der Baan, 1959) where shelducks have fed in this way. The birds may also dig in the mud for buried invertebrates, and can dig or paddle out craters 4–45 cm across and 2–12.5 cm deep, especially on the Grosser Knechtsand moulting area (Oelke, 1970, and my own observations). Incubating females, during their feeding bouts, and ducklings, particularly small ones, often feed by moving rapidly across the mud, pecking at individual items (figure 2.6), probably *Corophium* and *Nereis* (Buxton, 1975). On the Ythan a change from mainly sieving to almost exclusively pecking occurred among both males and females in early May (Buxton, 1975).

The different feeding techniques are generally used at different stages of tide. Bryant & Leng (1975) found head-dipping to be the commonest method, reaching a pronounced peak soon after the midpoint of the flow tide. Upending, understandably, was commonest in the deeper water around high tide, while those birds still feeding on the ebb and low tide used sieving on the mud surface or dabbling at the water's edge. Buxton (1975) made similar observations on the Ythan.

2.6 World distribution

The shelduck has a broad distribution, confined to the Palearctic, from western Ireland to the western parts of China. The latitudinal range, however, is fairly narrow, avoiding both the sub-arctic/boreal regions and most of the Mediterranean and southern Eurasia (figure 2.7). There is an interesting discontinuity in the range, with a coastal northwest European population in mild maritime areas almost completely separated from a predominantly inland eastern one occupying warm semi-arid areas (figure 2.7). Small local populations around the Mediterranean lie between the two main areas.

Almost half of the northwestern breeding stock occurs on the coasts of the British Isles (Atkinson-Willes & Scott, 1963). Shelducks also breed commonly all along the western seaboard of Europe, from the middle of Norway to the Gironde estuary in France, and at the western end of the Baltic with a recent extension to Finland.

Inland breeding appears to be increasing in Britain, Denmark and Germany, with the colonising of lakes, sewage works and large rivers. Small populations occur in a few places around the Mediterranean, with about 250 pairs breeding in the Camargue in southern France (Hafner, Johnson & Walmsley, 1979), and small numbers in Sardinia, Greece, northern Italy and Tunisia (Cramp & Simmons, 1977; figure 2.7). In winter there is a generally southward movement, with shelducks largely absent from Scandinavia and occurring in Spain and North Africa. In mid-winter there are large concentrations on the coasts of the Heligoland Bight, southwest Denmark, northern Holland and on larger estuaries of east and southeast England (Atkinson-Willes, 1969).

The eastern stock breeds from around the Black and Caspian seas in a narrow band, mainly between 40 and 50 degrees north, through the southern USSR (Dementiev & Gladkov, 1952), to the northwest of Manchuria and western China (Étchécopar & Hüe, 1978). Lönneberg

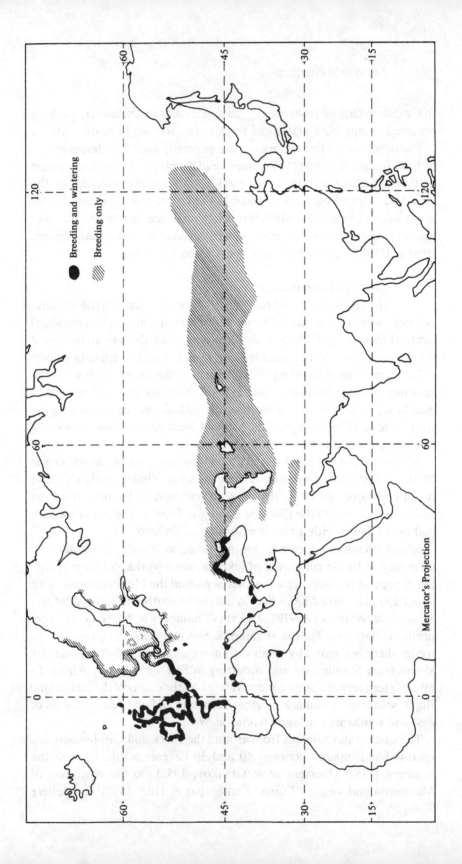

Breeding and wintering

Breeding only

Mercator's Projection

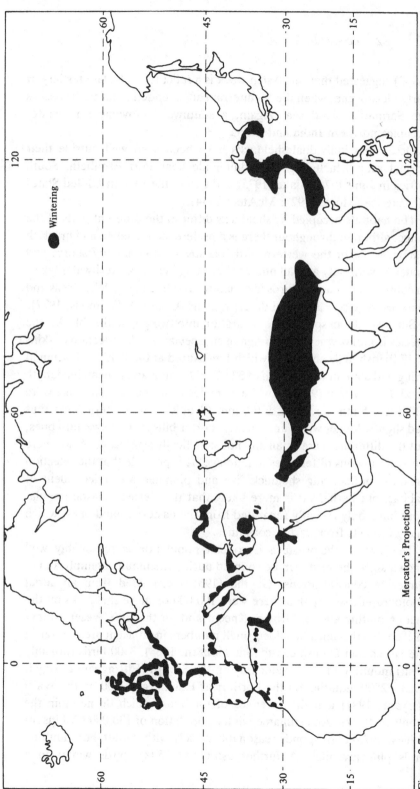

Figure 2.7 The world distribution of the shelduck. Occurrence is patchy within the indicated eastern range. Drawn from maps and data in Dementiev & Gladkov (1952), Hüe & Étchécopar (1970), Sharrock (1976), Cramp & Simmons (1977) and Johnsgard (1978).

(1932) suggested that this Asian stock is a relict from the late tertiary or early Pleistocene, when the Mediterranean extended as far to the east as the Sarmatic inland sea. Again, a southward movement in winter occupies northern India and China.

Occasional individual shelducks have been seen well outside their normal range, including one adult male near Port Elizabeth, South Africa in June 1974 (Blake, 1975) and two in the eastern United States of America (Morse, 1921; McAtee, 1944).

The habitats occupied by shelducks differ in the different parts of the range, although throughout there is a preference for saline and brackish waters. Most of the western birds occupy shallow muddy shores and estuaries with only a small number breeding inland on freshwater lakes. In contrast, the eastern birds live mainly on saline inland lakes, seas and marshes in steppe and semi-desert regions (Cramp & Simmons, 1977).

Surprisingly, in spite of the separation into two population blocks, the species is monotypic, with no sign of subspeciation. A collection of skins of 17 British male shelducks which I measured at the British Museum at Tring had a mean wing length of 324.7 ± 1.9 mm and a mean bill length of 53.9 ± 0.7 mm (9 bills), while 18 males from the eastern range were 330.8 ± 1.5 mm and 53.1 ± 0.6 mm (9 bills). The eastern birds thus had slightly longer wings and slightly shorter bills than the western ones, but the differences are small and not statistically significant. Also, there were no obvious differences in plumage. It is possible that the selection pressures influencing shelduck size and plumage are independent of habitat, but perhaps it is more likely that the western coastal populations have only recently separated from the eastern inland ones, which may be relicts from earlier coastlines.

The size of the western European population is reasonably well known since the birds can be counted during the moult assembly and in wintering concentrations. Goethe (1961a) estimated from an aerial photographic survey that there were 100 000 or more shelducks on the main moulting area of Grosser Knechtsand off the Elbe/Weser estuary in 1955. To this must be added small numbers in adjacent areas as far as the Dutch and Danish coasts (Salomonsen, 1968), 3400 birds (possibly Irish) moulting in Bridgwater Bay on the Severn (Eltringham & Boyd, 1963), 2500 moulting on the Forth near Edinburgh, 2000 on the Wash (Bryant, 1981) and the juveniles of the year, which do not join the moulting flocks. An estimated winter population of 130 000 (Atkinson-Willes, 1976) corresponds reasonably closely with the total of moulting birds plus juveniles. A further estimated 75 000 birds winter in a separate southern European area (Atkinson-Willes, 1976) but the size of the eastern population is unknown.

2.7 The shelduck in Britain
Numbers

Early accounts (e.g. Turner, 1544, in Saunders & Eagle-Clarke, 1927) mention the shelduck as a common resident in Britain, but in 1559 the sale of 'skeldraikis' was forbidden by an act of the Scottish parliament (Baxter & Rintoul, 1953), perhaps an early conservation measure! Certainly the British shelduck population seems to have declined by the late nineteenth century, through persecution and industrial or housing development in its feeding and nesting areas. Yarrell & Saunders (1885) considered that the increase in human population had affected the shelduck, although it still bred 'sparingly' in several coastal counties. In Ireland they stated that it had been 'driven by persecution, reclamation of land and other causes' from the former breeding areas, although it still bred where unmolested. Similarly, Seebohm (1885) found the species scarcer in southeast England than elsewhere and that, owing to persecution, it had become much rarer than formerly, being still most abundant 'in little-frequented places or in places where it is protected'. Several local bird faunas confirm this picture of decline. Thus Walpole-Bond (1938) described the shelduck as 'a not too common autumn-to-spring visitor' in Sussex prior to 1904, while in Kent up to 1909 only scattered pairs occurred from above Gravesend to the mouth of the Thames, with the species occurring only as an autumn and winter visitor elsewhere in the county (N. F. Ticehurst, 1909). Similarly in Suffolk the shelduck was a common breeder until around 1860, but had almost disappeared by 1890 (C. B. Ticehurst, 1932) through persecution prior to 1886 by local shooters who considered that the birds disturbed the rabbit warrens (Bannerman & Lodge, 1957)!

In the early twentieth century a general increase in the breeding population began, probably through increasing protection. In Sussex the first breeders in 1904 had increased to 500 pairs by 1938 (Walpole-Bond, 1938) although this was considered an over-estimate by Shrub (1979) who put the recent total at 100–150 pairs. In Kent, the low numbers in 1909 were followed by a 'wide extension of breeding range . . . (and) . . . enormous increase in total population during the nesting time' (J. M. Harrison, 1953) and in Suffolk a few pairs in the late nineteenth century had increased to 187 pairs by 1925–27 (C. B.

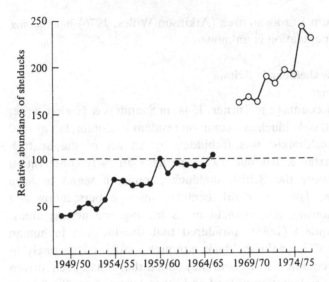

Figure 2.8 Changes in the relative index of abundance of shelducks in winter, relative to the count in 1959–60. Drawn from data in Harrison (1967) (up to 1966, closed circles), Bryant (1978) and Atkinson-Willes & Salmon (1977) (from 1969, open circles).

Ticehurst, 1932) and to 'at least five times' that total by 1960 (Payn, 1962). A similar story of recovery following an earlier decrease is mentioned in a number of other less complete county natural histories.

The increase in the British shelduck population has continued. Alexander & Lack (1944) referred to it as 'increasing throughout Britain' although they did not present any supporting data. J. Harrison (1967) showed that there had been a steady increase in winter numbers from 1950 to 1966, and this trend continued to at least 1977 (Atkinson-Willes & Salmon, 1977) (figure 2.8). The increase seems to have been general throughout Britain, with associated expansion in range (see below), although there have been some local decreases, probably through coastal disturbance, in south Wales, parts of Lancashire, Northumberland, Ayrshire and Loch Lomond (Parslow, 1967, 1973). Much of the general increase may have been by colonising (or recolonising) unoccupied areas, since several long studies of local populations have shown fluctuations but no systematic increase in numbers (see section 10.1). Since a recovery from earlier persecution should have been completed fairly quickly after the start of complete protection, the continuing increase is presumably a response to a gradual change in the environment, perhaps nutrient enrichment of estuaries which could have improved the food supply.

No very detailed census has been made of the present British shelduck population, although the approximate number of birds can be estimated. Using Goethe's (1961*a*) estimate of 100 000 moulting birds at Heligoland, and an estimate that roughly half of the birds ringed while moulting there were subsequently recovered in the British Isles, Atkinson-Willes & Scott (1963) estimated the population at approximately 50 000 birds. (Goethe (1961*b*) found that 55 per cent of 56 distant recoveries of moulting birds came from the British Isles, so the estimated total might perhaps be 5000 higher.) In addition, a further 8000 shelducks now moult in Britain (see section 2.8). An international wildfowl count in January 1967 gave a total of 113 500 shelducks in the Palearctic region, with 46 per cent of the birds in Britain (Atkinson-Willes, 1969). This gives an estimate for British shelducks of 45 000 which agrees fairly well with the estimate from the moulting population.

Yarker & Atkinson-Willes (1971) further estimated that breeding birds would form less than half of the total overwintering population, and suggested that there might be about 12 000 breeding pairs in the British Isles, or 10 000 pairs on mainland Britain. Sharrock (1976) considered that the shelduck densities encountered in a major survey of distribution were consistent with this estimate, which remains the best to date. However, the total should be treated cautiously since the estimates were taken from a wide spread of years during a period of increase, and the proportion of non-breeders in the population varies considerably, depending largely on breeding success in previous years (Patterson, Makepeace & Williams, in press *b*).

Distribution

The present British shelduck population is distributed widely around the coastline, wherever there are suitable shallow muddy or sandy shores with associated nesting habitat (Sharrock, 1976; figure 2.9). Density has increased especially in south and east England, and the range has extended to the southwest as far as the Scilly Isles. There has been an increase in the number of pairs breeding inland on lakes, reservoirs and sewage works, although most are still within 20 km of tidal water. The few exceptions to this are 30–32 km from the sea on the rivers Nene and Ouse in Cambridgeshire and Huntingdonshire, and one or two pairs over 100 km from the sea in Warwickshire (all the above distribution data from Sharrock, 1976).

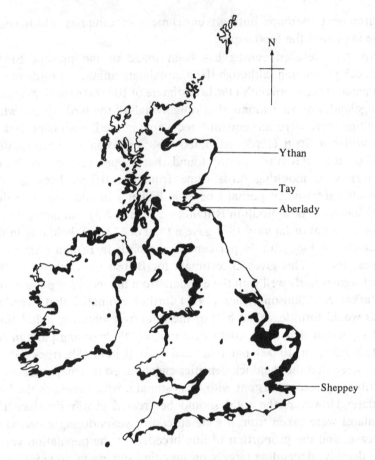

Figure 2.9 The breeding distribution of the shelduck in Britain (re-drawn from Sharrock, 1976). The study areas of four principal shelduck studies are shown.

2.8 The moult migration

Moulting areas

In common with the rest of the ducks, geese and swans (as well as auks, divers, rails and a few other groups), shelducks shed all their flight and tail feathers simultaneously and become flightless for several weeks in late summer. Like some of these other species, shelducks have a spectacular moult migration, first described by Hoogerheide & Kraak (1942), from all parts of their northwest European range to the Heligoland Bight or German Waddensea area just off the Weser and Elbe estuaries (figure 2.14). Here, vast areas of mudflat (especially near the Grosser Knechtsand), extending over 20 km offshore, afford protection against predation and have a good food supply, thus satisfying two of the main requirements of a major moulting ground (Salomonsen, 1968). Once flightless, shelducks become extremely wary (Goethe, 1961b) and flocks which I have approached on foot begin to run at the first sign of figures on the skyline up to 3 km away. Few of the breeding areas have large enough mud areas for such freedom of movement and the shelducks' plumage is too conspicuous to allow the birds to conceal themselves effectively in cover while moulting as many dabbling ducks do.

A small proportion of the European stock does, however, moult in more restricted areas, with up to 3400 in Bridgwater Bay on the Severn (Perrett 1951, 1953; Eltringham & Boyd, 1960, 1963; Eltringham, 1961) over 2500 on the Forth (Bryant & Waugh, 1976; Bryant, 1978) and over 2000 on the Wash (Bryant, 1981). All of these areas are said to have soft glutinous mud which restricts access by predators and the birds on the Forth 'were invariably found at the edge of the most remote mudflats on the estuary' (Bryant, 1978).

The moulting area in Bridgwater Bay has probably been used for at least 140 years. Smith (1844) reported that 'burrow ducks' were common in Somerset and that they collected at breeding areas '. . . from about May till July or August after which time they become more scarce in that immediate locality . . . but they are still to be found spread over the great expanse of mud and shallow water in the Bristol Channel throughout the year'. About 2000 birds were seen in Bridgwater Bay on 29 August 1926 and on 20 September 1936, and other large flocks were reported in August and September 1935, 1938 and 1949 (Report on Somerset Birds, 1926, 1935, 1936, 1938, 1949). Perrett (1951) saw 1800 birds in September 1950 but moulting was not confirmed until 1952 (Perrett, 1953). Aerial counts were made in 1959 and 1960 (Eltringham & Boyd, 1960, 1963) and counts from the ground from 1956 onwards (J. V. Morley and the Nature Conservancy Council, unpublished data). The peak number for the season, and the maximum proportion of birds moulting, both occur in early September (Eltringham & Boyd, 1963). The maximum numbers of shelducks seen in September have fluctuated considerably, and tended to be lower in 1960–74 than in 1958–9 and after 1975 (figure 2.10). It is possible that some of the birds were missed by the counters in some years, since J. V. Morley (*in litt.*) found that the ducks moved offshore from late July, presumably when they began to moult, and they were then much more

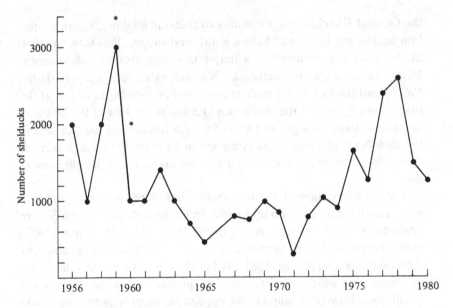

Figure 2.10 Maximum numbers of shelducks counted at Bridgwater Bay in September 1956–80. The asterisks show aerial counts (Eltringham & Boyd, 1963). The main data were supplied by J. V. Morley and the Nature Conservancy Council.

difficult to see. However, Eltringham & Boyd (1963) compared counts throughout the year and found aerial counts to be on average only 1.7 per cent lower than ground ones ($n = 4875$ birds). In June, aerial counts were 6.7 per cent lower ($n = 1310$) and in August 12.5 per cent higher than the corresponding ground counts ($n = 2000$). Unfortunately, no comparisons were made during the main moulting period in September, but in 1960 the maximum ground count for the month was only half of the maximum aerial count (figure 2.10). I visited the area on 1–5 September 1980 and found that in calm clear weather the whole bay could be surveyed, with gulls easily visible far out into the Bristol Channel, beyond the furthest visible shelducks. It is thus likely that if several counts were made each September including some in good weather, the maximum counts would probably be realistic estimates of the changes in numbers between years. The low counts in the 1960s and early 1970s are puzzling in view of the general steady increase in numbers of British shelducks. The change occurred too early to have been caused by the hard winter of 1962/63 and may show a decline in the local areas where the moulting birds originated or a period when more moulted near their breeding grounds or went on to Heligoland.

The moulting area in the Forth was first discovered in 1975 (Bryant & Waugh, 1976) and the number of shelducks using it has risen in parallel with the general increase in the British population (Atkinson-Willes & Salmon, 1977). The number using the Wash has also increased rapidly (Bryant, 1981). It is possible that an overall increase in the European stock may have led to the use of alternative sites to Heligoland (Bryant, 1978). Another possibility is that increased protection has meant less disturbance to the birds in these smaller estuaries. Certainly, in the past, even the shelducks moulting at Heligoland were not safe from human persecution. Goethe (1961*b*) described how large numbers of flightless birds were killed for meat and skins up till the early twentieth century. (Even following protection in 1934 the shelducks were not safe, since the use of the mudflats as a bombing range led to the death of over 12 000 birds in 1954. After this, however, the area was completely protected.) The rounding-up of the conspicuous flightless birds, as happened at Heligoland, may well have excluded moulting in any but the most extensive mudflat areas until recent protection. Bryant (1978) predicted an extension of local moulting flocks to some other major British estuaries and later found a flock of 2270 shelducks on the east Wash on 1 August 1980 (Bryant, 1981). If the risk of human predation is now small, the birds would benefit from moulting locally by avoiding the risks and energy loss of a long migration.

In the eastern part of the range, no single major moulting ground has been described, the birds apparently gathering in a number of smaller concentrations (Cramp & Simmons, 1977; Dementiev & Gladkov, 1952). For example, Hoogerheide & Kraak (1942) referred to a moulting area in the Volga basin and another on the Tibet–Sikkim border. This is consistent with Salomonsen's (1968) view that moult migration was not found in warmer areas, where most wildfowl moult in or near the breeding areas.

The outward flight

The departure on the moult migration begins in early July, with the yearlings and young adults leaving first, followed by adults which have failed in breeding (Lind, 1957). Adults with young delay their moult until the ducklings have fledged, although some males leave just before this. The behaviour of the departing birds has been studied mainly on the west coast of Britain, at Morecambe Bay (Coombes, 1949, 1950), the Cheshire Dee (Allen & Rutter, 1956, 1957, 1958) and the Severn (Morley, 1966), perhaps because the birds must head towards the land and thus be more conspicuous than east coast ones which depart over the sea.

All of the hundreds of departure flights seen were in the evening, almost all in the hour before and the hour after sunset (Coombes, 1950). Departures occurred only in fine weather, with a clear sky being essential and a following wind very important. No significant departures were seen during overcast weather or against an adverse wind. Flocks of up to about 100 shelducks rose calling from the shore and flew inland, although they usually zig-zagged back toward the sea several times while gaining height and some, or occasionally all, of the flock might break away and return to the shore. Most, however, continued eastward and the overland route has been confirmed by sightings of migrating flocks (Morley, 1966) and occasional birds on freshwater lakes inland (Coombes, 1950). A female shelduck ringed at Slimbridge on the Severn was recovered dead on the Wash on the east coast on a fairly direct line to Heligoland (Eltringham & Boyd, 1960). In 1956, 3000–4000 shelducks were seen passing eastward from the north Norfolk coast on 14–17 July (Allen & Rutter, 1957). It is not known whether the Morecambe and Dee birds stop at the east coast 190 km away, which they should reach just after midnight at their observed flying speed of 96 km per hour (91–8 km per hour in seven measurements by Morley, 1966). It is possible that they fly on directly to Heligoland, a further 560 km, which they should reach in the early morning. The birds seen departing from Morecambe Bay and the Dee had certainly assembled there from elsewhere, since the total seen leaving far exceeded the known summer populations of adults and yearlings (Coombes, 1950; Allen & Rutter, 1956). The additional birds may have come from Ireland or have gathered into pre-migratory flocks from adjacent breeding areas, perhaps because of a reluctance of individuals or small parties to risk an overland crossing. In Bridgwater Bay in July there is an influx of birds which later depart before the main moult flocks arrive (Eltringham & Boyd, 1963). The possibility of attack from birds of prey while the ducks were away from the refuge of water may have led to the migration taking place by night in fairly large groups, which could most easily be collected together from a big assembly.

Almost all of the very large number of departures seen (over 19 000 ducks from the three areas over several years) took place in July. Only in 1957 were some flights from the Cheshire Dee seen from 19 June and on 1 August (Allen & Rutter, 1958). The peak of departures from the west of England occurred on 18–20 July; on 18 July at Morecambe Bay in 1949 (Coombes, 1950), on 18 July from Bridgwater Bay in 1964 (Morley, 1966) and 19 July in 1955, and on 20 July in 1956 and 19 July in 1957 from the Cheshire Dee (Allen & Rutter, 1956, 1957, 1958). This shows remarkable consistency, given that the birds apparently did not leave in adverse conditions.

Arrival and the moult period

Shelducks begin to arrive on or near the main moulting grounds from the middle of June (Goethe, 1961b) and Lind (1957) showed movement through Jutland from then until the end of July, with a peak just before mid-July. Counts made around the Heligoland Bight itself are difficult to interpret, since large numbers of shelducks are known to move around within the area in response to tide, weather and human disturbance (Goethe, 1961b). However, numbers reached a peak off Trischen, 35 km northwest of the Knechtsand, around 20–25 July in 1950 and 1952 and off Mellum, 10 km southwest of the Knechtsand, around 20 July in 1954 and 1955 (Goethe, 1961b), as would be expected if most shelducks migrated at the same time as those seen leaving from the west of England.

In the main moulting area, the Grosser Knechtsand, large numbers of shelducks were present from about 20 July to the end of August, with the highest numbers counted in early August: on the fourteenth in 1964, the ninth in 1965, the fifth in 1966 and the eleventh in 1967 (Oelke, 1969a). During the moult period most of the birds moved within an area extending 20–30 km from the small sand islet of the Knechtsand, feeding on the mudflats at low tide and moving on the incoming tide to rest on exposed sandbanks at high water. Oelke (1974) marked five moulting shelducks with radio transmitters and found that they usually moved 3–8 km, sometimes up to 10 km, between feeding and resting places, appearing to drift more or less passively on the tide. Moulting shelducks spend much time preening, and feed for less of their time than birds that can still fly (Oelke, 1974). No precise measurements of the length of stay of individual birds have been made. Oelke (1969a), from changes in numbers and ringing recoveries, estimated that each bird stayed only 12–14 days. This seems remarkably short, since Hoogerheide & Hoogerheide (1958) found that shelducks remain flightless for 25–31 days.

The number of shelducks on the Grosser Knechtsand moulting area begins to drop from mid-August and few remain after mid-September (Oelke, 1969a). At the same time numbers rose in neighbouring areas,

Figure 2.11 Flocks of moulting shelducks on the Grosser Knechtsand. A few birds retain their flight feathers (bird in right foreground, lower figure) but most are flightless. Photos by Prof. Dr. H. Oelke.

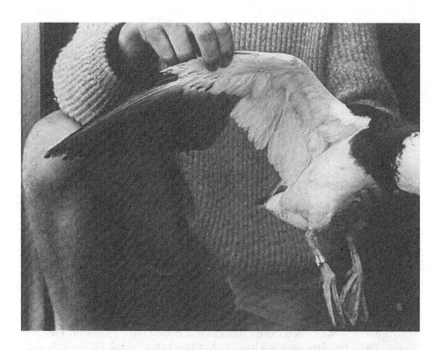

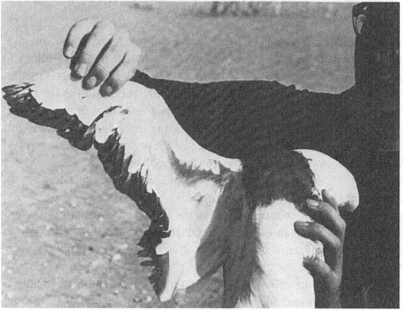

Figure 2.12 Shelduck with moulted secondaries, still retaining some old primaries (upper figure) and one growing new flight feathers. Photos by Prof. Dr. H. Oelke.

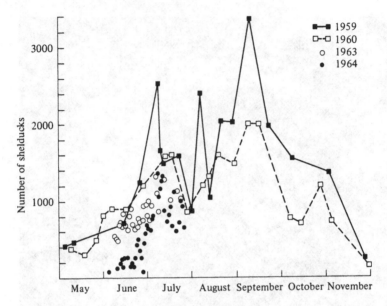

Figure 2.13 Counts of shelducks at Bridgwater Bay in summer and autumn. Re-drawn from Eltringham & Boyd (1963) (years 1959 & 1960) and data supplied by J. V. Morley (years 1963 & 1964).

with a peak off Sharhorn, 15 km north of the Knechtsand, in mid-September in 1955 and 1957. Shelducks also returned to Mellum in September in 1954 and 1955 after an absence in August (Goethe, 1961b). These observations suggest that only the initial phase of the moult occurs on the Knechtsand and that the birds then move outward, confirming Oelke's (1974) estimate of short individual stay. The outward movement continues through September, reaching the bays at Wilhelmshaven, 50 km southwest of the Knechtsand, in late September and early October (Goethe, 1961b) and extending into the return movement.

In the small British moulting areas the timing of the migration seems to be similar to that in Heligoland. On the Forth the birds concentrate into the moulting area in July, with most in full wing moult in early August, and start to spread out over the estuary again from mid-September (Bryant, 1978). At Bridgwater Bay, the number of shelducks rises to a peak in mid-July and then falls towards the end of the month (figure 2.13) as birds are seen leaving to the northeast, presumably towards Heligoland (Morley, 1966). Numbers then increase again to a higher peak in early September. Similar levels were seen in 1959, 1960,

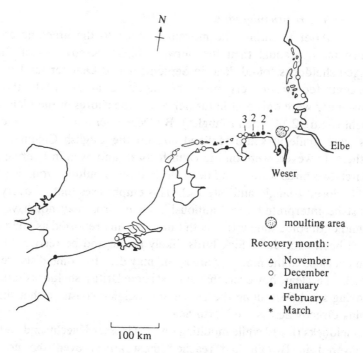

Figure 2.14 Recoveries of British-ringed shelducks on the European coast in autumn and winter. The numbers show several recoveries close together. Data supplied by the British Trust for Ornithology.

1963 and 1964, except for three very high counts in 1959 (figure 2.13). (It is noticeable that the counts in 1959, the first year of the aerial survey, were much more variable than in 1960. The high counts may include some duplication or may show real short-term fluctuation in numbers.) Numbers then decline steadily through late September, October and November.

Eltringham & Boyd (1963) found that the first birds became flightless (those which did not fly in response to the survey aircraft) in late July. Most of the birds present were flightless throughout August with a peak in the number flightless in early September; a majority were able to fly again by mid-September. In the first week of September 1980 I found that only nine per cent of the adults were flightless, with the rest having new primaries, so that the moult seemed to have been completed earlier than in 1959–60. The period of six to eight weeks when flightless birds are present corresponds with Hoogerheide & Hoogerheide's (1958) measurement of 25–31 days of flightlessness in seven captive birds, assuming some variation of timing within the population.

The return migration

After moulting, the movement back to the breeding areas is much more gradual than the outward flight. Recoveries of British-ringed shelducks found dead in September and October tended to be concentrated in or very near the moulting areas, while those in November were only a little further round the shores of the Heligoland Bight (figure 2.14, open triangles). By December and January recoveries were made in south Holland and on the English Channel coast. Others, however, were on the Danish coast, and even in February and March two birds were found dead close to the moulting grounds (figure 2.14, closed triangle and asterisk). This emphasises that recovery data must be interpreted very cautiously; some birds may not have been found until some time after death (although any reported as 'long dead' have been excluded). Sick birds, likely to die and be recovered, may remain close to the moulting areas and may drift for some distance after death. It is clear, however, that at least some British shelducks return by moving southwest along the Dutch and Belgian coasts, rather than by flying directly across the North Sea.

Shelducks ringed while moulting on the Grosser Knechtsand and later recovered in Britain had reached most areas, even the north of Scotland, by December and were well distributed by March (figure 2.15). Winter recoveries showed some concentration in southeast England, with 45.9 per cent of the 74 recoveries in October to March being between the Wash and Kent, while only 32.1 per cent of the 28 birds found from April to July were there (figure 2.15). This suggests that the returning migrants first return to the southeast and may spend some time there before dispersing. It is also possible that some continental shelducks winter in southeast England. Since much mortality occurs then, particularly in hard weather, any such overwintering stocks would give a concentration of recoveries. By the breeding season (April–July, figure 2.15), recoveries were distributed round Britain, with 35 per cent of them on the west, including one in Ireland. In contrast with this, 48 out of 50 shelducks ringed in Britain and later recovered in the moulting area originated from the east coast. This, however, is highly misleading and certainly chiefly reflects the distribution of trapping and ringing rather than the distribution of migrant shelduck. Almost half of these recovered birds were ringed in the small population of the Ythan Estuary, Aberdeenshire, which makes up less than one per cent of the estimated British population!

Shelduck ringed on the Ythan show a distribution of recoveries rather

Recovery months:
* October – December
o January – March
● April – July

100 km

Figure 2.15 Recoveries in Britain of shelducks ringed while moulting on the Grosser Knechtsand, Heligoland. The numbers denote several recoveries from the same place. Data supplied by the British Trust for Ornithology.

different to that of the British population as a whole. Apart from the expected concentration around the moulting area, most recoveries were in northern Britain, with only four on the Channel coast and southeast England (figure 2.16). Scottish shelducks may possibly cross the southern North Sea to north England or at least not stay long in the southeast. Five recoveries on the west coast and to the north of the Ythan, unlikely to be part of the normal return migration route, may represent birds

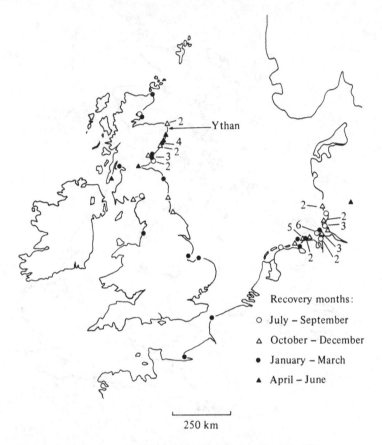

Figure 2.16 Recoveries of adult shelducks ringed on the Ythan Estuary, Aberdeenshire. The numbers denote several recoveries at the same place. Data supplied to the ringers by the British Trust for Ornithology.

which joined flocks returning to other areas, or may be individuals originally ringed on transient visits to the Ythan.

Winter counts of shelduck on major British estuaries show large wintering concentrations which later disappear. Hori (1964*a*) found 'large numbers of winter visitors' with February flocks of up to 1400 on the Swale Channel, Sheppey, while the local breeding population was only 65–147 pairs (Hori, 1969). Similarly, in Langstone Harbour near Portsmouth, January or February peaks of 1508–3130 birds were reduced to 345–640 by April in four winters (Tubbs, 1977). Undoubtedly some of the wintering birds must have dispersed later to local breeding areas but only a small proportion can be accounted for by this.

Young (1964a) showed from Wildfowl Trust counts that the winter peak number occurred progressively later in more northerly estuaries, with peak numbers at the Stour in Essex in December; at Egypt Bay, Kent and Breydon Water, Norfolk in January; at Holbeach, Lincolnshire, and Teesmouth and Fenham Flats, Northumberland in February, and at Edenmouth in Fife in March, suggesting a steady return movement from one major estuary to the next.

Smaller estuaries generally do not have wintering populations larger than the breeding numbers. At Aberlady Bay on the Firth of Forth Jenkins (1972) found that numbers reached approximately the spring level of about 120 birds by January in 1969–71 but did not exceed it, whereas at the nearby large estuary at Grangemouth, December to February maxima of up to 1562 birds were reduced to usually under 300 by March. On the Tay estuary (Boase, 1951) and on the Ythan (Patterson *et al.*, 1974) winter numbers were low, the winter assembly for these and probably other populations in northeast Scotland most likely being the Eden estuary in Fife, where I have seen several colour-marked shelducks from the Ythan in the winter flock of 1200–1500 in January and February.

It is very unlikely that small estuaries could support winter populations much larger than the breeding ones, whereas on large estuaries the breeding populations may well be restricted by lack of nesting habitat to levels well below those which could be supported by their vast mudflats. The larger and more variable feeding areas of large mudflats may also be more reliable in winter, when cold weather may make other food supplies inaccessible. Thus a number of influences may have led to the existing pattern of winter assembly in a series of major estuaries, followed by dispersal to breeding areas.

2.9 The annual cycle

On their return from the moult migration to the breeding area, the adult shelducks first form loose flocks, where the birds feed gregariously. However, as early as January, on warm sunny days some pairs leave the flock and space themselves out along the shore for a few hours. This territorial behaviour becomes general by mid-April, and by early May all the pairs which will have territories that season have taken them up and stay on them throughout the day (Williams, 1973). A proportion of the population, mainly one to three-year-old birds which have recently arrived, do not become territorial and remain in a flock confined to any areas not defended by the territory owners.

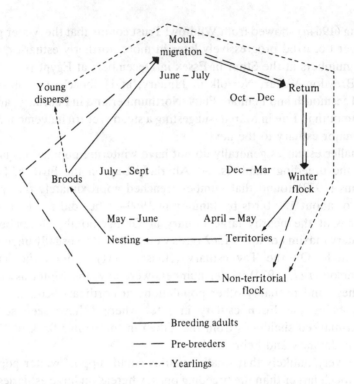

Figure 2.17 The annual cycle of the shelduck, showing the
approximate season for each phase. At about three years of age
pre-breeders join the territorial component of the population.
Throughout the nesting and brood period, adults whose breeding
attempt fails join the non-territorial flock.

At the same time as pairs take up territories they begin to prospect for
nest sites, flying from the shore to adjacent nesting areas soon after
dawn. Here they again become gregarious as the pairs aggregate into
small groups, called 'parliaments' (Young, 1970*a*) or 'communes' (Hori,
1964*a*) where the birds interact, display and search for nest sites.

While the female lays and incubates her eggs the male spends most of
his time on the territory, although he may rest a short distance from the
nesting burrow for short periods. The female leaves the nest three or
four times each day, always during daylight, and feeds voraciously on
the territory for one to one-and-a-half hours each time. The male
defends her while she feeds, doing much of his eating while she is away.
When the female returns to the nest he escorts her and then returns to
the territory, usually without alighting.

When the young hatch they are led to the shore by both parents but are not usually taken to the territory. Instead the social organisation changes again, to one of overlapping ranges occupied by the different broods in any area. The parents, however, defend an area around their young wherever they go within the range, leading to considerable aggressive interaction between brood adults and to consequent mixing and creching of broods.

During the period of brood-rearing, the flock of young adults, augmented by failed breeders, begins to decline in size as the first birds leave on their moult migration. By late July, only the parents and their ducklings remain. As the young reach fledging, at seven or eight weeks of age, the parents migrate, often with the male leaving first, while the young join with other broods to form small groups which wander around the breeding area and then quickly disperse for their first winter.

The social organisation of shelduck populations thus changes markedly with the seasons, from large assemblies in wintering estuaries to smaller winter flocks in breeding areas, to spaced territories (with gregarious grouping in nesting areas at the same season), to overlapping ranges with defence around the brood and finally to flocks again during the moult. At several points during the season, the breeding adults, the young pre-breeders and the juveniles of the year are distributed separately (figure 2.17). The following chapters will explore behavioural and population questions raised at each stage of this complex annual cycle.

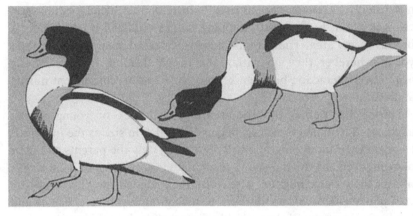

3

Gregariousness: the winter flock

Grouping together into flocks, herds or shoals may benefit individual animals in a number of ways, such as reducing the risk of predation or increasing the efficiency of prey capture (Bertram, 1978). However, the proximity of other animals must intensify competition for resources such as food or mates. Conflict between individuals frequently results in a high level of agonistic display and threat, or even fighting, and the development of dominance hierarchies in the groups.

In this chapter I will describe agonistic behaviour and dominance in flocks of shelducks and will discuss the characteristics which lead to some individuals becoming relatively dominant. I will consider the benefits obtained by the dominants and the selection pressures which have favoured the striving for dominance. Finally, I will discuss the consequences of dominance on the limitation of winter population size.

3.1 Arrival

On leaving their last major winter assembly estuary, shelducks begin to gather in their breeding areas. Some, resident in a major estuary system such as the Thames (Hori, 1964a), merely remain behind

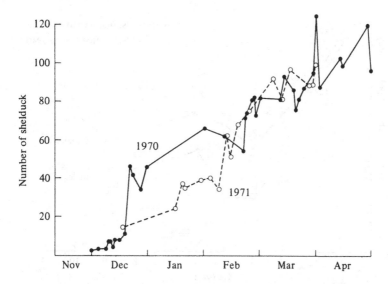

Figure 3.1 The number of shelduck counted on the Ythan estuary through the winter in 1970 (from Patterson, 1977; closed symbols) and 1971 (open symbols).

when the mass of migrants leaves; most, however, return to a small estuary or stretch of muddy shore.

The time of return varies from place to place and between years. At Aberlady in 1969–71, shelduck numbers began to rise rapidly in early October and reached the level of the breeding population by early January, while in other similarly sized populations in the same area, arrivals started only in January and continued until April (Jenkins, 1972). In earlier years (1949–57) at Aberlady the increase started later, beginning only in early November and not reaching breeding numbers until February (Jenkins, 1972). Progressive arrivals, starting in January on the Tay and in February at Montrose, were recorded by Boase (1951, 1959). On the Ythan, the first birds usually arrive in late November or early December and the steady increase common to most of the breeding areas again continues till April (figure 3.1; Williams, 1973; Young, 1964a).

It is not clear why different populations, often quite close together, should arrive at different times. Jenkins (1972) considered that disturbance, particularly from shooting of other wildfowl, might deter shelducks. Arrival has occurred earlier at Aberlady in recent years, coinciding with the establishment of a nature reserve, but remained later at nearby Tyninghame Bay where there was considerable shooting.

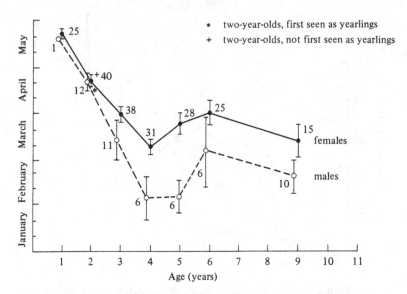

Figure 3.2 Mean dates when ringed shelducks of known age were first seen on the Ythan. The sample sizes are shown beside the points.

Similarly, at Tullibody, further up the Forth, where shooting was extensive, high numbers of shelducks were not recorded until a month or more after the end of the shooting season (Jenkins, 1972).

The likelihood of mudflats becoming frozen, so that food becomes unavailable, may also influence the date of arrival on the breeding area, particularly towards the north of the shelduck's range. On the Ythan, winter numbers often drop during periods of freezing weather in mid-winter. Two birds caught and ringed on this estuary in winter were recovered soon afterwards at Montrose, 76 km to the south, suggesting that the birds may move back to larger estuaries in hard weather. There is considerable evidence that shelducks suffer heavy mortality during freezing weather (e.g. in the winter of 1962–3; Dobinson & Richards, 1964; Harrison & Hudson, 1964) which might make it risky for them to arrive too early in small, less saline estuaries.

The ability to cope with difficult feeding conditions may be one reason why older females arrived on the Ythan significantly earlier than younger ones. Ythan birds ringed as ducklings showed a progressively earlier arrival with increasing age, up to four years old (figure 3.2). Older birds showed no further tendency to arrive earlier; the slightly later mean arrival dates were due to some birds not being seen until very late in the season, possibly having been missed earlier. After the first two years, males mostly arrived earlier than females (figure 3.2). Two-year-old females which had been seen as yearlings arrived only slightly earlier than those which had not been seen in the first year.

The earliest arrivals were not necessarily those which departed first at the end of the previous season. Indeed, many of the first arrivals were old males which had bred successfully the year before, delaying their departure while caring for ducklings, so that they had completed their moult migration much more rapidly than birds which had not reared young. Evans & Pienkowski (1982) found that, at Aberlady, birds which had reared young generally arrived later and had more variable arrival dates than other adults. Possibly such experienced old birds, the selected survivors of their year-classes, had a detailed knowledge of the feeding areas, which enabled them to forage more efficiently in difficult conditions and allowed them to risk moving north earlier. The advantage in doing so may lie in earlier attainment of a territory and breeding; Williams (1973) showed that pairs seen together earliest in the season were usually the first to take up territories, and that the earliest pairs on territory were generally the first to breed.

Since male shelducks on average arrived earlier than females, the first flocks seen on the Ythan each season had a marked predominance of males (table 3.1) with the sex ratio approaching equality only in April

Table 3.1. *Sex ratio in the winter flock on the Ythan, 1980*

	Total examined	Percentage of males
1–15 January	12	100.0
16–31 January	18	77.8
February	115	60.9
March	192	62.5

and May. Some mated pairs were first seen together, but in a majority the male was seen before the female (table 3.2). Possibly the males, being larger, are able to cope better with an unpredictable winter food supply but their earlier arrival may also reflect a selection pressure on the males to get a territory and breed as early as possible.

Table 3.2. *Relative arrival dates of the male and female in Ythan ringed territorial pairs, 1978*

	before female	Male first seen same day	after female
Number of pairs	20	6	5
Percentage of pairs	64.5	19.4	16.1
Mean number of days	41.9	—	10.6
Standard error	6.4	—	2.9

3.2 Dispersion

The term 'winter flock' may be misleading since feeding shelducks usually form only loose flocks and the birds of a breeding area may split into several sub-flocks. Hori (1964a) described two main winter aggregations at Sheppey, and Jenkins *et al.* (1975) referred to 'one or more flocks' at Aberlady. The Ythan birds occurred on several areas of the estuary in winter, although most were seen on the larger mudflats in the middle reaches (figure 3.3) where disturbance was minimal. Buxton (1975) found surprisingly that the birds did not spend most times in areas of highest food abundance. Site 5 (figure 3.3) was little used in 1971 when it was popular with wildfowlers but was occupied commonly in 1972 when shooting had declined (Williams, 1973). The first birds to arrive used only the lower and middle estuarine sites (sites 1 & 2, figure 3.3) in December and early January, possibly because these were less likely to freeze than the upper reaches. Sites 1 to 5 were used in January and February, but sites 6–10 in the upper estuary were used only in March and April as the flock began to disperse and pairs moved closer to their territory positions (Williams, 1973).

Individual shelduck usually winter in the same sub-flock. Williams (1973) found that among 89 ringed birds, each seen at least three times, 30 per cent were always at the same site, and 71 per cent had two-thirds of their sightings at one place. Each bird generally also returned in subsequent years to the same wintering site, which was usually close to the position of the bird's territory. The small proportion of individuals which were seen equally commonly at various sites throughout the estuary were mainly non-territorial or had territories on fresh water away from the estuary later in the season (Williams, 1973).

Within the main parts of the winter flock, the birds stay close

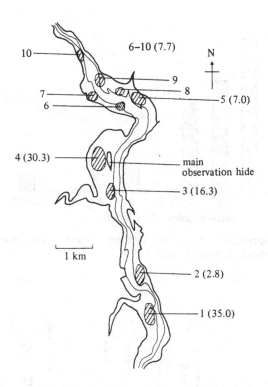

Figure 3.3 Sites on the Ythan estuary used by the flock in January to March 1971. The bracketed figures are the percentages of flock birds counted at each site (total = 988 counted on 22 occasions). Modified from Williams (1973).

together. Spacing was measured on the Ythan by estimating the distance between each male and its nearest neighbouring male. Female to male spacings were avoided since the smallest distances would merely reflect the closeness of mated pairs. The distances could be estimated easily to the nearest metre up to 10.0 m between birds, to 2.5 m up to 15 m but only to the nearest 5.0 m for birds further apart. Most males had their nearest neighbour at two or three metres and few were more than five metres away from others (figure 3.4). The mean nearest-neighbour distance remained between two and six metres throughout the winter and showed only a very slight and non-significant tendency for the distance to increase through the period (figure 3.5). Although individual pairs left the flock altogether to take up territory from mid-March onward, the remaining flock birds did not space further apart.

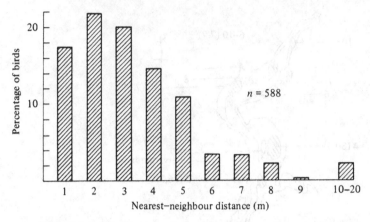

Figure 3.4 Distribution of nearest-neighbour distances between males in the winter flock, January to March 1972.

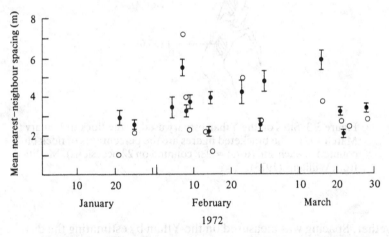

Figure 3.5 Mean spacing between nearest neighbours throughout winter 1972. The filled symbols show the mean spacing on each day and the vertical bars indicate standard errors. The open symbols show mean spacings between birds which subsequently interacted. From Patterson (1977).

3.3 Feeding

More than half of a shelduck's time in winter is spent feeding. On the Forth an average of 57 per cent of the birds seen were foraging. The proportion varied through the tidal cycle, with over 90 per cent feeding at the flow tide but only nine per cent at low water (Bryant & Leng, 1975, mean calculated from data in their table 1). Buxton

Table 3.3. *Proportion of birds feeding at any one time throughout the tidal cycle, Ythan estuary 1973. Modified from Buxton (1975)*

| | Males | | Females | |
	Number of observations	Percentage of birds feeding	Number of observations	Percentage of birds feeding
22 February	729	62.7	399	63.4
8 March	429	26.8	306	35.0
15 March	983	10.3	788	14.9
Total	2141	31.4	1493	32.0

(1975) found that at one main site on the Ythan, 63 per cent of the birds were feeding (averaged over the tidal cycle) in February, but that this proportion dropped to 10–15 per cent by mid-March (table 3.3). Evans & Pienkowski (1982) found a similar drop at Aberlady. The drop was somewhat steeper in males than in females. The abundance of the main prey species, *Hydrobia, Corophium* and *Nereis,* did not increase over the winter (Buxton, 1975) and the decrease in time spent feeding probably represented time being given to other activities, particularly aggression, although some birds (about five per cent) were seen feeding on newly sown grain in fields in March.

3.4 Agonistic behaviour

Shelducks are noticeably aggressive birds, and threat, attack, avoidance behaviour, and even fights, are commonly seen among the winter flock birds. Hori (1964a) found that display and interaction between pairs at Sheppey were common in February and led to interacting groups of up to 50 birds.

The simplest form of threat by both sexes is a direct lunge. made towards the opponent, by extending the neck with the bill slightly open. This movement can become a physical attack if the victim is close enough to be pecked and does not move quickly enough.

In a more ritualised threat, Head-down, the back feathers are raised to give a hump-backed appearance, and the head and neck are held out parallel to the ground and pointing towards the opponent (figure 3.6). Boase (1935), who described this posture as 'the usual act of menace', noted that the head may be moved up and down or from side to side, but such movements are not conspicuous. In the Head-down the bird may remain stationary or may walk, swim, run or even fly at an opponent. In the flying variant, the head may be held below the level of the belly.

Figure 3.6 Head-down posture. Drawn from a photograph.

Figure 3.7 Head-throwing by the male and Inciting by the female.
Drawn from a photograph.

The most conspicuous display, given almost exclusively by the male, is Head-throwing (Hori, 1964a), also called Bowing by Boase (1935). In this, the head is drawn back then thrown rapidly upwards and forwards. The movement is repeated so that the head describes a circle over the shoulders (figure 3.7). The intensity of the movement varies considerably, from a few perfunctory nods of the head to a very intense form in which the head makes large circles and the forepart of the body rises and falls in time to the movement. The display is always accompanied by rhythmic and often loud whistling, and is commonest in prolonged interactions between closely matched males or between pairs.

The corresponding female display is Inciting, described as a scooping action by Boase (1935), where the head and neck are extended parallel to the ground and the head then moved, often vigorously, from side to side (figure 3.7). The female gives a loud harsh barking quack during this

Figure 3.8 Alert posture. Drawn from photographs.

display, which is performed while she stands beside or runs around her mate, who usually responds by Head-throwing and whistling. I have often had the impression that aggressive interactions were initiated by females running up to their mates and Inciting, with the male subsequently making an attack on another male or pair nearby, but perhaps some less conspicuous action by one of the males occurred first.

Avoidance of threat or attack sometimes involves only a mild moving away in a resting posture, but birds subject to vigorous aggression often show an Alert posture with the head high and the plumage sleeked (figure 3.8). A similar posture, perhaps less intense, has been described by Boase (1935) as Rest-intent, by Johnsgard (1965) as the High-and-erect, and by Jenkins *et al.* (1975) as Alert. In this the head is held higher and the neck straighter than in the resting posture and the black feathers of the nape appear to be raised. On land, birds in this posture may stand with the front part of the body raised so that the breastband and belly-stripe are visible from the front (figure 3.8). Undoubtedly a bird in this posture is well placed to be watchful, but it is difficult to decide whether disturbances, potential predators or other shelducks are the object of attention. Boase (1935) described his Rest-intent as occurring 'when excitement is rising in a group', and Jenkins *et al.* (1975) considered that the frequency of Alert changed when the risk of predation did not. They showed that the proportion of birds showing Alert varied with tidal state so as to be highest when few were feeding and also high when the birds were crowded together at freshwater drinking places, suggesting that the posture is related to interaction between shelducks. Possibly closer study would show two separate but similar postures, one with sleeked head feathers shown by birds which are disturbed, fearful or being attacked by other shelduck, and one with nape feathers raised, used in social situations where the bird is less likely to flee and where the posture may indicate awareness of a potential opponent.

Shelducks of both sexes which are about to fly up in alarm may perform Bill-tipping, described as the Salute by Boase (1935). The head and neck are held straight up in the Alert and the bill tip is jerked sharply upward, often to the extent that it points vertically up for an instant. The movement may be repeated several times until the birds fly or relax again. I have found when attempting to approach shelducks that Bill-tipping is a useful warning of imminent departure! The movement is used mainly when birds are alarmed by people or potential predators, but can also occur in males about to fly after being attacked and may act as a pre-flight signal between mates.

A rather puzzling movement, the Whistle-shake (Johnsgard, 1965), was also described by Boase (1935) who did not name it. In this, given by males, the body is reared up and the head and bill pointed upwards while the whole plumage is given a rotary shake, to the accompaniment of a trilling whistle (figure 3.9). The bill is then lowered on to the breast as the body is returned to the resting position. This movement is shown in apparently mildly alarming situations, as when a person appears at a distance, but may also be given by a male which has just been attacked or threatened by another.

Figure 3.9 Whistle-shake. Drawn from a photograph.

The various agonistic postures of the shelduck are not equally common. In 818 encounters seen in early 1970 on the Ythan, 48 per cent involved running at the opponent in the Head-down, with a simple lunge being the next commonest form of attack at 29.1 per cent of those

Table 3.4. *Relative frequency of the different agonistic postures, Ythan estuary, 1970*

Posture		Percentage of interactions
None		2.6
Lunge		29.1
Head-down	walk	8.6
	run	48.0
	fly	6.7
Head-throw/Incite		3.4
Unknown		1.6
Total number of interactions		818

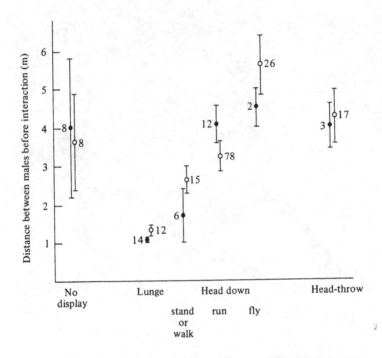

Figure 3.10 Distances between opponents using different threat postures. Closed symbols refer to interactions in January and February, open ones to those in March (1972). Vertical bars are standard errors and sample sizes are shown alongside.

seen (table 3.4). In 2.6 per cent of incidents one male supplanted another without any obvious posturing, and only 3.4 per cent of the interactions involved Head-throwing.

The different postures were used in slightly different circumstances, especially when opponents were at different distances. The lunging attack was used only when opponents were close, while walking, running and flying in the Head-down were used with increasingly distant opponents (figure 3.10). Head-throwing was shown between fairly distant opponents, and interactions with no display occurred at a wide variety of distances.

Most of the agonistic encounters seen on the Ythan were simple supplanting movements, when one bird threatened or attacked another which immediately moved out of the way. The attacker sometimes, but by no means always, then occupied the spot where the victim had been feeding or resting. Rarely (see section 3.6) a victim stayed put or even threatened back. In these cases the attacker then usually withdrew, but

Figure 3.11 Two male shelducks in the early stages of a fight. Drawn from a photograph.

the encounter could (rarely) escalate to physical attack or fighting between the birds (figure 3.11). Encounters between two pairs often became prolonged and complex, with vigorous Head-throwing by the males and Inciting by the females.

A large majority of encounters were between males, and they made 76.7 per cent of the attacks seen. Females attacked much less frequently, making only 17.4 per cent of attacks, most of which were against other females (table 3.5). The small number of individual females seen to attack males or pairs almost always had their mate displaying nearby at the time. Pairs interacted chiefly with other pairs.

Table 3.5. *Distribution of aggressive interactions between the sexes on the Ythan, 1970. Data for 1971 were very similar. From Patterson (1977)*

Attacker	Male	Victim Female	Pair	Total
Male	**548**	55	24	627
Female	34	102	6	142
Pair	4	0	45	49
Total	586	157	75	818

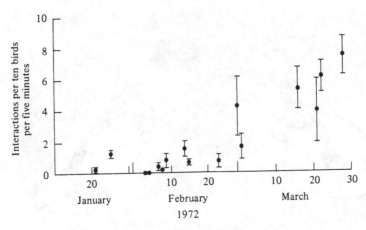

Figure 3.12 Seasonal change in the frequency of aggressive interaction in the winter flock. The vertical bars show standard errors around day means. Re-drawn from Patterson (1977).

3.5 Frequency of aggressive interaction

Although aggression is always noticeable in shelduck flocks, its frequency of occurrence varies and can be difficult to measure. For example, Jenkins *et al.* (1975), selecting particular individuals to watch and recording their behaviour every second, found that aggression occurred in only 0.0–0.6 per cent of the observations in mid-winter and reached 1.0–5.0 per cent in only four out of 14 watches. To overcome this low rate of occurrence in any particular individual, I selected groups of ten birds on the Ythan and watched the whole group for five minutes. The groups (often including several obvious pairs) were taken randomly from the flock, avoiding only the extreme edges. During the five-minute observation some of the original ten birds might wander away from the others, but new ones were incorporated so that a constant ten were maintained under observation, albeit with a slowly changing membership. The number of threats and attacks among the ten birds was counted by watching from a hide or car at 100–150 m and the observation was discarded if the birds became alarmed during the watch.

The frequency of agonistic interaction remained very low from late January (when observations began) until the end of February and then increased throughout March (figure 3.12). There also seemed to be an increase in the intensity of the encounters; the proportion preceded by running or flying, rather than by standing still or walking, increased from 56 per cent (of 66) in January and February to 86 per cent (of 163) in March (χ^2 = 22.2, P < 0.001). There seemed to be a similar increase in the amount of pecking and physical combat, but this was difficult to measure. Jenkins *et al.* (1975) also found that the proportion of time birds spent in Alert, which they considered to be associated with social interaction, increased from a maximum of 15.3 per cent of observations in winter to 58.2 per cent in spring.

Why should shelducks become more aggressive towards the end of the winter period? What advantage does a male obtain by increasingly attacking others? The defence of feeding space as more birds arrive in the breeding area can probably be excluded. Numbers on the Ythan rose steadily throughout January and February (figure 3.1) without a corresponding increase in aggression, and at Aberlady the number of birds remained stable through the winter with no increase in spring (Jenkins, 1972). The spacing between neighbouring birds on the Ythan did not change significantly over the winter (figure 3.5), the flocks presumably spreading out as more birds arrived.

I attempted to estimate the distance between interacting birds, prior to the interaction, by monitoring their positions continuously and by noticing impending attacks in their early stages as birds converged, watched each other or displayed. However, birds which subsequently interacted were not consistently closer together than others (figure 3.5), so we must look for reasons other than defence of individual feeding space to explain the increasing aggression.

There is no evidence that food became less abundant in late winter. Buxton (1975) measured the density and biomass of the three commonest food items in the Ythan flock areas in 1971 and 1972 and showed that food abundance did not change over the winter. The birds also spent less of their time feeding in March (table 3.3). Low temperatures and especially freezing of the mud may make the food less available (Goss-Custard, 1969) and large numbers of shelduck have been found dead in emaciated condition in hard weather (Dobinson & Richards, 1964; Harrison & Hudson, 1964). However, interaction frequency on the Ythan was lowest in the coldest part of the winter and highest in March (figure 3.5) when temperatures were higher. It thus seems unlikely that males were interacting over a scarcer food supply.

The progressive increase in aggression in March could be seen as a transition to the high levels shown in territorial defence later in the season. The change from flock to territory is gradual and pairs can be seen visiting their territory site for increasing periods over a number of days or weeks. Since the males, especially, are much more aggressive on territory than in early winter (Williams, 1973), it is possible that their involvement with their territories carries over into an increased aggression when they return to the flock. Also, as the breeding season approaches, it may well be in a male's interests to defend his mate more vigorously to prevent her being annexed by another male. The pressure from unmated males may also increase towards late winter as they attempt to get a mate. Hori (1964a) attributed much of the interaction in late winter flocks to attempts by unmated males to break existing pair bonds and to harassing of established pairs by groups of unpaired males. Thus, the individual advantage of increased aggression in late winter would seem to be concerned with the attainment of a breeding territory and the formation and maintenance of a pair bond, rather than with the acquisition of food or feeding space.

The consequence of such behaviour for the population is most likely to be limitation of the number present in winter, if some individuals are excluded, or are deterred from settling, by the aggression of others. Jenkins et al. (1975) suggested, mainly from counts of birds and frequency of Alert posture, that aggressive behaviour in winter at Aberlady was important in limiting the size of the subsequent breeding population. They suggested that areas of good feeding are limited, that there should be a dominance hierarchy in the winter flock and that subordinate birds would be excluded from the best feeding areas and eventually from breeding. To test these ideas it is necessary to find whether consistent dominance relationships exist in wild shelducks in winter, whether such behaviour could exclude subordinate individuals from feeding areas and whether dominance could affect breeding and survival.

3.6 Dominance

The first indication that some shelducks might be able to dominate others is that any individual initiating a threat or attack is very likely to win the interaction. In 137 encounters between 11 shelducks I kept in captivity, only twice (1.5 per cent) did the victim retaliate. Similarly, in the Ythan winter flock in 1972, retaliation occurred in only three (1.3 per cent) of 228 interactions seen between males. Gilboa (unpublished) found an even lower percentage (0.9 per cent) of retaliations among 1494 interactions seen in 1978. This might suggest that the birds attacked only those subordinate to themselves. However, other explanations are possible; for example, if there is little worth defending retaliation might not be worth the risks involved, or the birds might assess each other's motivation so that a satiated bird would not contest a feeding site against a determined hungry one. Such alternatives to the existence of consistent dominance between particular individuals can most readily be excluded by observations on marked birds. These of course must first be caught.

Trapping shelducks

Surprisingly, in spite of their predominantly invertebrate diet, shelducks are attracted to grain. Both barley and wheat, usually soaked, have been used on the Ythan, and wheat was effective at Aberlady (Jenkins *et al.*, 1975). In both areas the birds quickly learned to come to feed at a baiting point, allowing traps of various kinds to be used.

The commonest trap used at the Ythan was a simple funnel design, triangular in plan with a funnel in one (4 m) side and a door in another. The 1.8 m high walls were made of 50 mm mesh plastic-covered chain-link fencing; the covering and the diamond-shaped holes minimising damage to the shelducks' soft bills. The roof was of thinner-gauge wire netting, raised by a pole in the centre where a small hole allowed the birds to escape if the tide rose exceptionally far. The trap was sited on a mudflat used by the flock with the funnel on the side facing towards the land, since birds trying to escape tend to head towards the water. Despite this precaution it became obvious when watching the trap that many shelducks learned to find the funnel, so that they were able to enter, feed and leave again at will. To prevent this, I fitted a set of vertically hanging wires which opened inwards but not outwards. This modification greatly increased the total catches although not the number of new unringed birds being caught; the wires may have deterred inexperienced birds from entering as well as preventing experienced ones from leaving. Other similar though more complex traps were used by Young (1964b) and Williams (1973).

A proportion of shelducks, although feeding on grain around traps, would not enter them. Some of these birds were caught in a clap net, operated by two sets of powerful springs. These pulled on cables around pulleys at the bases of 1.8 m aluminium alloy poles pivoted in two steel frames. Bait was scattered near the net in its set position and the poles were released from a hide when a group of shelducks had gathered at the bait. A similar technique using a cannon-driven net was used at Aberlady (Evans & Pienkowski, pers. comm.).

Jenkins *et al.* (1975) used the narcotic drug tribromoethanol (Avertin) on wheat to catch 51 birds in 1972–73. The main problem with this method is the difficulty of controlling how much each individual will eat, with the consequent risk that some birds, perhaps some already marked on a previous occasion, may be killed.

Small numbers of adult shelducks have been caught by other methods. Young (1964a) caught a few territorial birds by setting up a small funnel trap containing a decoy near the middle of the territory. Hori (1964a) caught some females on the nest, but Young using the same technique found that many of the birds deserted the nest afterwards.

Young shelducks can be caught before they can fly, either by running them down on mudflats (Young, 1964a), by tracking them to hiding places in reedbeds (Clapham & Stjernstedt, 1960) or by driving them into corrals of netting (Patterson *et al.*, 1974). On the narrow estuary of the Ythan, a 'V' of netting was erected with the point of the 'V' on the water's edge at low tide and pointing downstream. A 100 m length of 1.3 m high netting extended into the stream while another 0.6 m high length reached to the top of the shore. The 'V' opened into a small catching pen where the ducklings were held after being very carefully eased downstream by about three people. Several broods of almost-fledged young could be caught at once and seemed to sort themselves out afterwards, since each set of parents regained the same number of ducklings they had before catching.

Marking shelducks as individuals

Shelducks are easy to recognise by coloured leg rings since they have long tarsi capable of holding several rings, and they spend a large proportion of time out of water with their legs visible, in contrast to many other duck species. Early studies used celluloid rings (Young, 1964a) but these faded badly and later workers have used Darvic (Imperial Chemical Industries), a rigid P.V.C. very resistant to fading. Using five distinct colours and four coloured rings on each bird (one above a numbered metal ring on one leg and three colours on the other), a very large number of birds can be recognised individually. If rings are replaced when they become worn, marked shelducks can remain recognisable for long periods; one Ythan male marked in 1962 was still identifiable in 1979. Colour combinations can be identified at ranges of up to 400 m with a good telescope.

Ducklings present more of a problem since five rings would be an excessive load, and on the Ythan less than a fifth of the young ever returned to the local population, so that the use of colour combinations on them seemed unjustified. Instead, Ythan ducklings were given a single tall ring engraved with a number. The colour of the ring and the leg it was on defined the year-class of the duckling, and the number identified the individual. The main disadvantage of engraved rings was that they were more difficult to read than colour combinations and could not usually be identified beyond 150–200 m. When ducklings were recaptured in later years their engraved rings were replaced by conventional colour rings.

Plumage dyes can be used on the extensive white areas of the adult shelducks and enable them to be identified at ranges of up to two kilometres while swimming and while in long vegetation. Hori (1964*a*) found that alcohol-soluble dyes were quickly washed off by the birds, but *p*-phenylenediamine (black) and picric acid (yellow) have given good results on the Ythan, providing marks which lasted from mid-winter until the body moult in July. Five positions along the neck and flank were used (Patterson, 1979) to give 30 combinations in each colour for either sex.

Using these different marking techniques, up to 85 per cent of the Ythan population was marked for individual recognition and in some years up to a third of these were dyed as well as ringed. This gave a sound basis for investigation of dominance relationships.

Dominance relationships

I watched the Ythan flock from a hide (figure 3.3), using grain to attract the birds close enough to identify their rings easily. The bait also increased both the density of the flock and the frequency of agonistic interaction so that data could be collected more quickly than from the normally dispersed flock. Interactions between the same males seen at the bait and away from it gave the same results. Females interacted too seldom to be studied in detail.

The winner and loser of each encounter were readily identified since, as already mentioned, the victim of an attack very rarely retaliated. With enough data it was possible to see whether the relationship between any two males was consistent from one meeting to the next. In 82.4 per cent of 74 pairs of males, each pair having had at least six interactions, one of the males was always the winner. In these, the result ($> 6{:}0$) differed significantly from a hypothesis of equality between the two birds ($P < 0.05$, binomial test). In the remaining 13 pairs of males, the overall loser won only a single interaction in nine cases and the result differed significantly from equality in seven cases (table 3.6).

Table 3.6. *Proportion of encounters won by the overall winner in pairs of males. Each line represents one pair of males. From Patterson (1977) with 4 pairs from A. Gilboa labelled* a

Encounters won by		
overall winner	overall loser	P (Binomial test)
30	1	<0.01
10	1	<0.01
10[a]	1	<0.01
9	1	<0.01
8[a]	1	0.017
7	1	0.04
5	1	
5	1	
5	1	
12	4	0.04
7[a]	4	
6[a]	2	
5	5	

There was clearly a high degree of consistency in the results of interactions between any two marked birds, especially as any mistakes in the identification of rings would tend to lower this consistency. Male shelducks thus seem to form stable dominance relationships and this explains why victims of attacks rarely retaliated; they were presumably attacked only by birds dominant to them. Since a large proportion of relationships were entirely one-way, dominance between male shelducks can be predicted from a single encounter with a high probability. This makes it possible to describe the dominance hierarchy in a wild population, where it is difficult to record large numbers of encounters between every possible pair of individuals.

Hierarchies among captive shelduck

I made preliminary observations on two groups of captive shelducks kept in 10.0 × 10.0 m enclosures with small pools. Aggressive encounters between them were watched for 30 min after the renewal of their food, when there was most interaction around their trough.

A group of four adult males, kept with an adult female and four

Table 3.7. *Dominance hierarchy among four captive male shelducks, November 1968. Each number represents the encounters won by the bird in that row over the bird in that column. From Patterson (1977)*

Winner	A	B	Loser C	D	Total won
A	—	9	6	8	23
B		—	11	2	13
C			—	4	4
D				—	0
Total lost	0	9	17	14	40

hand-reared juveniles in 1968, showed a clear straight-line hierarchy (table 3.7) with no retaliation by subordinates. In the enclosure the juveniles clearly dominated the adults, but this may not be general since the young birds were more familiar with the situation than were the adults, which were often clearly ill at ease while being observed. A second group, of nine hand-reared juvenile males, kept with three juvenile females in 1970, showed very similar results.

Hierarchies in wild shelduck

Ideally wild birds could be ranked in the same way as captives, by inspection of the results of encounters between each pair of marked individuals. In practice, however, it was not possible to see every marked Ythan male interacting with every other one, since only a selection of them was at the observation point at any one time and some came rarely (see below). A male could only be placed within the hierarchy if it was seen interacting with a number of others which were themselves distributed through the rank order. For example, if a bird was defeated by two low-ranking males it was clearly also subordinate but if it was only seen to be beaten by the top-ranking male it could have been placed anywhere in the hierarchy.

An alternative ranking method is based on each bird's success in all its encounters, including those with unmarked opponents and those with other marked birds which had not themselves enough data to be included in the hierarchy. For each male, the percentage of its encounters which it wins gives a measure of its relative dominance. A disadvantage of this method is that different males interact with a different selection of opponents.

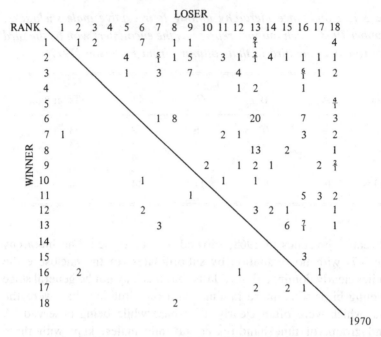

Figure 3.13 Dominance table for 1970. The birds are numbered in rank order along the rows and columns. Each figure in the table shows the number of encounters won by the bird in that row over the bird in that column. Where a bird won only a majority of the encounters, the proportion won by each is shown in the winner's row. From Patterson (1977).

Data on the interactions between the marked birds were sorted, stored and analysed by computer at the Aberdeen University Computing Centre, and a preliminary ranking of the males was based on the percentage of their interactions won against all opponents. This first rank was then modified by inspection of the observed dominance relationships between pairs of individuals until an order was achieved which gave the smallest number of inconsistencies.

In both 1970 and 1971 it was possible to arrive at a rank order (figure 3.13) with a small number of inconsistent results (where males defeated others which apparently were higher in the rank order). Some of these, and some of the retaliations by subordinates (proportional results in figure 3.13) may have been due to mistaken identities. Most, however, probably reflect a normal lack of complete consistency in a large group of wild birds which are not in close contact all the time. Very similar results were obtained in 1971 and by A. Gilboa in 1978 and D. J. Tozer in 1979 (Patterson *et al.*, in press *b*) but much less consistent results were obtained at Aberlady (Evans & Pienkowski, 1982).

Usually not all of the individuals in a wild bird population can be assigned a dominance rank. The most obvious exceptions are the unmarked birds which can rarely be recognised individually by the observer. Since most of the ringed birds were caught in winter in baited traps they could have been a biased sample: either the weaker, hungrier members of the flock, or conversely the more dominant ones which were able to drive others from the food.

Differences in average dominance between marked and unmarked birds can be measured by considering all the unmarked ones together in their encounters with all the marked birds. If there is no bias in the sample of marked birds, they should win around 50 per cent of their encounters with a large sample of unmarked ones. On the Ythan, marked males won 43.0 per cent of 272 encounters with unmarked birds in 1970, and 52.7 per cent of 201 encounters in 1971. These percentages do not differ much or consistently from 50 per cent, so there is no suggestion that the marked sample was biased with respect to dominance. Gilboa (unpublished), however, found that marked males won 69.6 per cent of 749 interactions with unmarked males in 1978. Williams (1973) has shown that marked and unmarked birds in this population did not differ significantly in breeding success; in the three years 1970–2 taken together, when 21.7 per cent of all 189 territorial (breeding) pairs were unmarked, 17.4 per cent of the 69 pairs which hatched ducklings and 14.3 per cent of the 21 pairs which fledged young were also unmarked. Thus, unmarked pairs were somewhat, but not significantly, less well represented among successful pairs than in the breeding population as a whole.

A second important group which cannot be assigned a rank are those marked birds which are seen interacting with insufficient different opponents to allow their position in the hierarchy to be defined. Such marked but unranked birds can, like the unmarked ones, be compared as a group with the ranked birds. This has been done for the Ythan population, using marked males which subsequently become territorial on the estuary in the same season so that the ranked and unranked birds were equivalent sets of breeding adults.

Few interactions were seen between ranked males and unranked ones, but the unranked birds were generally subordinate, winning only 21.2 per cent of encounters in 1970, 12.5 per cent in 1971 and 25.6 per cent in 1978 (table 3.8). The differences from 50 per cent (expected if the two

Table 3.8. *Interactions between ranked and unranked males. Each line represents one pair of marked males. From Patterson (1977), with 1978 data from A. Gilboa*

	Encounters won by			Encounters won by	
	ranked	unranked		ranked	unranked
1970	4	0	1971	11	0
	3	0		1	0
	3	0		1	0
	3	0		1	0
	2	0		0	1
	0	1		0	1
	0	1	Total encounters won	14	2 (12.5%)
	0	2	Total dominant	4	2
Total encounters won	15	4 (21.1%)			
Total dominant	5	3	1978	11	0
				3	0
				3	0
				2	0
				2	0
				2	0
				1	0
				4	2
				1	3
				0	1
				0	1
				0	3
			Total encounters won	29	10 (25.6%)
			Total dominant	8	4

categories were on average equal in dominance) cannot be tested statistically since repeated interactions between any two individuals are not independent of each other. However, considering each pair of males as one result, more ranked birds dominated unranked ones than the reverse, although the differences from equality in these few data were not significant in any year or with all years combined.

It is clear from these results that the shelduck winter flock is socially structured, with an obvious dominance hierarchy, at least among the males. Many of them can be assigned a rank order and those that cannot, because they take part in fewer interactions (at least at observation points), appear to be subordinate to the ones that can be ranked.

Changes in dominance with time

A group of nine hand-reared juvenile males, kept with three juvenile females, was observed three times to check for changes in the hierarchy. On 23 February, 1970, all but one of the males could be

Table 3.9. *Changes in relative dominance among nine hand-reared juvenile male shelducks, 1970. From Patterson (1977)*

26 February	9 March	26 March
A	E (+a♀)	E (+a♀)
B (+b♀)	A	B (+b♀)
C	B (+b♀)	I (+c♀)
D	C	A
E	D	F
F	I	C
G	F	D
H	G	G
	H	H
I (not ranked)		

ranked (table 3.9). The second-ranking male appeared to have a pair bond with the second-ranking female, since the two stayed together consistently. By 9 March, bird E, previously fifth in rank, had paired with the dominant female and moved into top place. The other males remained in the same rank order except that the previously unranked bird (I) was associating with the remaining female and had moved into third place in the hierarchy, so that the three paired males were in the top positions, with the original top-ranking male A in fourth place and F in fifth (table 3.9). The three top pairs were in order of the females' original relative dominance.

The hierarchy among these captive juveniles, although clear at any one time, was thus very changeable. Most of the upward changes, however, were associated with the formation of pair bonds, and possibly resulted from an increased tendency for paired males to attack others coming close to their mates in the confined space of the enclosure. It is interesting that the females did not pair with the original top-ranking males but that they subsequently seemed to increase their mates' status.

Among the Ythan wild population, nine males which could be ranked in both 1970 and 1971 changed their positions considerably and there was no significant correlation between their relative rank positions in the two years (figure 3.14). The higher-ranking males in 1970 tended to remain high in 1971 but the middle-ranking birds in 1970 tended to lose rank relative to those originally below them. These changes in relative rank were not related to changes of mate or territorial status or with

RANK

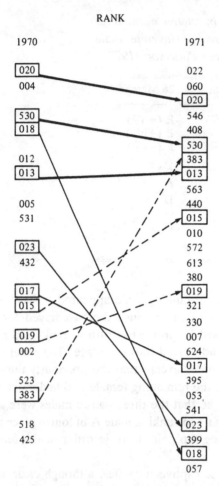

Figure 3.14 Changes in rank order between years. The figures are the serial numbers of individual marked males arranged in rank order. Birds which were ranked in both years are joined by arrows. From Patterson (1977).

breeding success in the intervening period. However, the two males which dropped most in relative dominance (023 and 018) were both unpaired in 1971. Although another male (530), also unpaired in 1971, remained high in rank, these results again suggest a relationship between pair formation and status. In spite of these year-to-year changes, relative dominance of males seems to remain unchanged within one season; the results of interactions seen later in the year, e.g. on the nesting grounds, were consistent with those seen in winter between any given pair of males.

Table 3.10. *Weights of ranked and unranked territorial males in winter (January–March). Data for 1970 and 1971 from Patterson (1977), for 1978 from A. Gilboa and for 1979 from D. J. Tozer*

Year and category	Number of birds	Range of weights (g)	Mean weight (g)	Standard error	d	p
1970 ranked	13	1150–1550	1319	35		
unranked	17	960–1500	1201	30	2.565	<0.02
1971 ranked	14	1120–1400	1268	20		
unranked	10	980–1300	1205	28	2.360	<0.02
1978 ranked	15	990–1360	1212			
unranked	5	1140–1360	1218			
1979 ranked	7	1070–1420	1228			
unranked	8	1160–1300	1231			

Individual characteristics associated with dominance

In many animal species, an individual's dominance is correlated with its physical or behavioural characteristics. For example, larger animals commonly dominate smaller ones, as in dairy cattle (Schein & Fohrman, 1955), gaur *Bos gaurus* (Schaller, 1967), voles and mice (Grant, 1970), giant tortoises *Geochelone elephantopus* (MacFarland, 1972), spiny lobsters *Jasus lalandei* (Fielder 1965) and others. Similarly, older black grouse *Lyrurus tetrix* tend to dominate younger ones (Johnsgard, 1967) and larger families of white-fronted geese *Anser albifrons* dominate smaller families, pairs and singles (Boyd, 1953).

The linear size of the Ythan shelducks was not measured, but body weights, which reflect both condition (especially the amount of stored fat) and underlying linear size, were obtained each winter when the birds were trapped for ringing.

There was no correlation between a male shelduck's weight and his position in the hierarchy in 1970 or 1971, or in similar data collected by A. Gilboa (unpublished) in 1978 and by D. J. Tozer (unpublished) in 1979 (18, 28, 42 and 35 ranked birds respectively). Comparison of the ranked males with the unranked ones (which tended to be subordinate as mentioned above) gave equivocal results. In 1970 and 1971 I found that the unranked males were significantly lighter than the ranked ones (table 3.10), but in 1978 and 1979 A. Gilboa and D. J. Tozer found no difference between the two categories. Shelduck thus seem to differ from many other species in having no strong or consistent tendency for heavier animals to be dominant.

Table 3.11. *Ages of ranked and unranked males.*
Data for 1970 and 1971 from Patterson (1977)
(augmented by data on birds other than those
territorial on the estuary), for 1978 from A. Gilboa
and for 1979 from D. J. Tozer. All the samples
exclude birds caught that season

| | Minimum age in years | | |
	over 5	not over 5	Fisher test
1970 ranked males	12	3	
unranked males	5	8	$p = 0.027$
1971 ranked males	12	13	
unranked males	1	12	$p = 0.012$
1978 ranked males	18	14	
unranked males	18	7	$p = 0.106$
1979 ranked males	16	16	
unranked males	19	15	$p = 0.174$

Only the minimum age of most of the ringed Ythan males was known, since they were already in full adult plumage (at least two years old) when first trapped. Only a minority were caught as juveniles or yearlings and were of known age. In 1970 and 1971 I found no tendency for older males to be higher ranking than younger ones but many of the birds had first been caught and marked only a few years before and were certainly much older than their minimum ages suggested. However, new data from A. Gilboa in 1978 and D. J. Tozer in 1979, when the oldest birds were over ten years old, also showed no correlation between dominance rank and age.

When ranked Ythan males were compared with the (more subordinate) unranked ones, there was no consistent difference in age. In 1970 and 1971, confining analysis to males which later became territorial on the estuary, I found that the ranked males were older than the unranked ones and were significantly so in 1970 (Patterson, 1977). This difference persisted when subsequent non-territorial males were included, with many of the ranked males, but fewer of the unranked ones, being over five years of age (table 3.11). However, the unranked males in 1978 and 1979 had a similar age distribution to that of the ranked birds. Thus there is no consistent tendency for older shelduck to dominate younger ones in the winter flock.

Similarly, in 1970 and 1971 there was no correlation between rank and either date of arrival on the estuary or breeding success in the previous season, so that early-arriving and successful birds did not gain any dominance advantage.

If the increased level of interaction in March (figure 3.12) is considered to be an early stage of territorial aggression, a male in a flock close to his eventual territory might tend to initiate more attacks than other males and might tend to dominate them. Brown (1963) showed that Steller jays *Cyanocitta s. stelleri* were more dominant when close to their territories than when further away, and Leyhausen (1956, 1971) found that domestic cats *Felis domesticus* were relatively more dominant near their personal sleeping places. The distance between the observation area and the centre of the eventual territory was measured for all the ranked Ythan males which became territorial. In 1971 there was a (non-significant) tendency for the males with territory sites closer to the observation point to be dominant to those with more distant territories. There was no such trend in 1970 or in more limited data from 1972. A. Gilboa (unpublished) confirmed this lack of correlation in 1978 by observing interactions at eight sites scattered along the estuary (rather than the single site used in 1970–2) to give a better distribution among the eventual territory positions. Six of the observation points were each visited by at least four males which were ranked and later territorial, but at none of the sites was dominance rank correlated with distance to the eventual territory. Usually each male occurred at only one of the sites but the few pairs of males which were seen at more than one site did not change their relative rank when moving from the principal site of one male to that of the other. The closeness of a shelduck male to his eventual territory thus does not appear to influence his dominance. The increase in the amount of aggressive interaction in late winter is thus perhaps unlikely to have been caused by increasing territorial aggression as I have suggested previously (Patterson, 1977), otherwise those males nearest to their territories might be expected to be dominant.

Since a male shelduck's status apparently cannot be predicted from simple physical characteristics, previous history or proximity to his territory, what then decides which birds will become relatively dominant? The finding that my captive juvenile males rose in rank after pairing (table 3.9) suggested that the presence of a mate might be associated with dominance. This was confirmed by A. Gilboa (unpublished) who found in 1978 that males in the upper part of the hierarchy were significantly more likely to have their mate present in the winter

Table 3.12. *Presence of mate and dominance in the winter flock. Dominance data for 1978 from A. Gilboa*

Part of rank order	Mate in flock present	absent	Fisher test		Mate in flock present	absent	Fisher test
1970 upper	9	0		1970 ranked	16	2	
lower	7	2	p = 0.235	unranked	5	18	p < 0.001
1970 upper	11	3		1971 ranked	17	11	
lower	6	8	p = 0.051	unranked	3	13	p = 0.007
1978 upper	20	1		1978 ranked	31	11	
lower	11	10	p = 0.002	unranked	14	26	p < 0.001

flock than were lower-ranking males. All but one of the more dominant birds were paired whereas almost half of the lower-ranking birds were still unmated (table 3.12). Unranked males (which are mostly subordinate to ranked ones) were also significantly less likely to be paired than were ranked males. Data from 1970 and 1971 showed a similar picture. In 1970 all but two of the ranked males were paired, the two unpaired ones being ranked 14 and 17 out of 18. Otherwise the results were very similar to those in 1978.

Thus, the most consistent correlate of dominance in male shelducks on the Ythan was the presence of the mate in the winter flock. This suggests that much of the observed aggression might be stimulated by the female. I have indeed seen many interactions which started by the female Inciting around her mate, who began to whistle and display by Head-throwing and then attacked another nearby male. Variations between the males in their aggressiveness in the presence of their mate would lead to a hierarchy, in which rank was not necessarily related to any physical characteristics of the birds.

An alternative explanation of the association between dominance and being paired is of course that females might prefer more dominant males. Even if dominance cannot be predicted from physical characteristics by the scientist, this does not exclude the possibility that the female might detect dominance, perhaps by watching a male's encounters with other males or by assessing his ability to defend her. Indeed dominant males may be more often paired through a better ability to retain their mate. However, the observations on my captive young shelducks showed that the females did not pair with the males that were initially most dominant and that the paired males subsequently in-

Table 3.13. *Territoriality in ranked and unranked males. Ranking data for 1978 from A. Gilboa and for 1979 from D. J. Tozer*

	Territorial	Non-territorial	Fisher test		Territorial	Non-territorial	Fisher test
1970 ranked	18	0	$p < 0.001$	1978 ranked	27	15	$p = 0.026$
unranked	8	15		unranked	17	23	
1971 ranked	23	5	$p = 0.066$	1979 ranked	16	19	$p = 0.147$
unranked	10	7		unranked	16	26	

creased in rank. This suggests that pairing leads to dominance rather than the reverse.

Dominance and subsequent performance

Being dominant can confer many benefits to fitness by giving preferential access to resources such as food or mates. Dominant woodpigeons have higher feeding rates than subordinate ones (Murton, Isaacson & Westwood, 1966) and in many species the dominant males sire the majority of the offspring, e.g. in laboratory mice *Mus* (DeFries & McClearn, 1970), Norway rats *Rattus norvegicus* (Calhoun, 1962) and black grouse (Johnsgard, 1967).

After leaving the winter flock shelducks take up territories, but there was no tendency in 1970 and 1971 for the more dominant Ythan birds to become territorial earlier than the subordinates. In 1970 all of the ranked males took up territories, while in 1971 the six which did not had a mean rank position of 14.8 (ranks 6, 8, 9, 19, 23 and 24) out of 28, showing that dominant males were not more successful in gaining breeding territories. Fewer of the ranked males obtained territories in 1978 and 1979 than in the earlier two years and there was a tendency for the more dominant males to be more successful, although the differences were not statistically significant. The ranked males were more successful than the more subordinate unranked ones in gaining territories (table 3.13), although the differences were significant only in 1970 and 1978.

Every year some of the Ythan territories were held on freshwater pools peripheral to the estuary and it seemed possible that subordinate birds excluded from the estuary might be using these sites. However, in 1970, relatively high-ranking males occupied territories on pools while in other years males of all ranks or mainly low-ranking ones occupied

Table 3.14. *Dominance rank of males with territories on pools peripheral to the estuary. Dominance data for 1978 from A. Gilboa and for 1979 from D. J. Tozer. Data for 1970 and 1971 from Patterson (1977)*

Year	Ranks of males with territories on pools	Mean rank	Total number of ranked males
1970	1, 4, 6, 8	4.8	18
1971	3, 12, 17, 22, 27	16.2	28
1978	17, 21, 24, 25, 40	25.4	42
1979	16, 32	24.0	35

them (table 3.14). In each year more of the ranked males occupied freshwater pool territories compared to the unranked birds, although the proportion doing so was small in both categories and the differences were not statistically significant. Dominant shelduck males thus seemed to have no consistent advantage in obtaining a territory and if anything, slightly more of them took territories off the estuary. Evans & Pienkowski (1982) found similar results at Aberlady.

An important potential benefit of dominance is an increased breeding output. The nesting success of territorial shelduck males can be measured by the proportion seen with newly hatched ducklings. (Frequent interchange of young between broods makes it impossible to determine how many of each pair's own young reach independence; see section 7.6.) Dominant Ythan males were not significantly different from subordinate ones in the proportion of them which were seen with a brood of ducklings (table 3.15). There was no significant or consistent difference in mean brood size between the two dominance categories.

In an earlier paper (Patterson, 1977) I showed that the unranked Ythan males were less successful than the ranked ones at hatching broods. However, re-analysis of the data showed that many of the unranked birds had only been seen interacting with others after the end of March, when the flock had begun to disperse. Since such late interactions in the remaining part of the flock might not be comparable with the majority, only birds seen interacting before the end of March were included in a new analysis of the 1970 and 1971 data, together with additional material from 1978 and 1979. There was still a trend towards more ranked territorial males being seen with ducklings, although the proportion was significantly higher only in 1979 (table 3.16). There was no consistent difference in mean brood size between the two categories.

Table 3.15 *Nesting success in relation to dominance rank of territorial males. Data for 1978 from A. Gilboa and for 1979 from D. J. Tozer*

Half of rank order	Number of males seen with brood	not seen with brood	Fisher test	Mean brood size	Standard error
1970 upper	4	5	$p = 0.242$	7.0	0.4
lower	6	3		8.2	1.0
1971 upper	6	6	$p = 0.267$	6.3	1.5
lower	4	7		5.8	1.1
1978 upper	6	11	$p = 0.317$	6.0	0.8
lower	3	7		10.3	3.9
1979 upper	4	5	$p = 0.343$	9.3	0.7
lower	4	3		8.5	1.6

Table 3.16. *Nesting success of ranked and unranked territorial males. Data for 1978 from A. Gilboa and for 1979 from D. J. Tozer*

Year and category	Number of males seen with brood	not seen with brood	Fisher test	Mean brood size	Standard error
1970 ranked	10	8	$p = 0.236$	7.7	0.7
unranked	3	5		8.0	1.3
1971 ranked	10	13	$p = 0.240$	6.1	1.0
unranked	3	7		6.3	0.7
1978 ranked	9	18	$p = 0.215$	7.4	1.6
unranked	4	13		4.3	0.7
1979 ranked	8	8	$p = 0.007$	8.9	0.9
unranked	1	15		(3)	—

The relationship between dominance and hatching success is obviously not clear-cut. The most dominant males were not more successful than those ranked just below them, but there seems some tendency for the most subordinate (unranked) males to be least successful. Evans & Pienkowski (1982) found no relationship between dominance and hatching success at Aberlady.

The influence of dominance on access to food is difficult to assess in shelducks, which normally feed dispersed over mudflats and rarely

Table 3.17. *The number of observation days on which individual marked males were seen at the baiting station. All had at least ten possible sightings. Data from Patterson (1977)*

Number of days seen	Ranked males	Unranked males
0		2
1		5
2	1	1
3		1
4	1	
5	1	
6	3	
7	1	1
8	3	
9	2	
10	2	
11	1	
Mean number of days seen	7.3	1.7
Mann-Whitney U = 20, $p < 0.002$		

appear to compete aggressively over food items or restricted feeding sites. However, the small baiting area used in the Ythan was apparently attractive to the birds, which crowded together to feed on it, giving the possibility that some birds might be deterred or excluded from feeding there. Most of the ranked males were seen almost every observation day, so that variances in attendance were minor. However, among birds first seen before February, with at least ten possible sightings at the bait, ranked males were seen on significantly more days than unranked birds (table 3.17). Away from the baiting area, the unranked males were seen almost as frequently as the ranked ones and were distributed similarly over the estuary. The eventual territory sites of the two categories also had similar distributions. Most unranked males were seen at the baiting site only once or twice out of the ten possible days and they had significantly fewer recorded interactions than ranked birds which were seen on the same number of days (table 3.18). This last difference is of course not surprising, since the ranked birds were selected as those which interacted with several other marked birds, and so would be expected to have had more interactions. The low number of interactions by the unranked birds could have resulted from their avoidance of encounters or merely by their being present at the bait for only a short time on each occasion. The two possibilities cannot be distinguished since it was not possible to record the duration of individuals' visits to the bait.

Table 3.18. *Number of aggressive interactions recorded for each marked male. Data from Patterson (1977)*

Birds seen once at bait		Birds seen twice at bait	
ranked males	unranked males	ranked males	unranked males
56	12	27	9
52	9	13	8
34	6	11	5
22	6		1
	4		
	3		
	2		
	2		
	1		
	0		
	0		
	0		
Mean 41.0	3.8	17.0	5.8
Mann-Whitney U = 0,		Mann-Whitney U = 0,	
$p = 0.028$		$p = 0.028$	

Since I have already shown that dominant shelduck males were not consistently heavier than subordinates, any preferential access to the best feeding sites does not appear to affect physical condition. This was perhaps not surprising in 1971 which was mild from January to March (mean daily maximum temperature near the Ythan in February was 7.7 °C and the mean minimum 0.7 °C). However, there was also no weight advantage to dominants in 1978 which was colder (February mean daily maximum and minimum 3.4 °C and −2.7 °C), presumably increasing any competition for food.

Any overall benefit of dominance to the fitness of the adult shelduck should be revealed by improved survival. However, only a small number disappeared from one year to the next and these were not consistently high or low ranking birds (table 3.19). Among unranked birds the number disappearing was also small, and the proportion was not consistently different from that among the ranked males.

Since subsequent performance was not clearly correlated with rank, what selection pressures then favour the apparent striving for dominance observed among male shelducks? Since the presence of a mate was associated with being dominant, the main selection favouring male aggression and dominance may well be the maintenance of the pair bond. Males which fail to pair, or which lose their mates to other males, will be unable to breed since male shelducks do not normally take up

Table 3.19. *Dominance ranks of males which disappeared, presumed dead. Data for 1970 and 1971 from Patterson (1977), dominance data for 1978 from A. Gilboa*

Year	Ranks of males which disappeared	Mean rank	Total number of ranked males
1970	2, 7, 14, 15, 17	11.0	18
1971	1, 24	12.5	28
1978	16, 20, 42	26.0	42

territories alone (see section 4.4). The increase in aggression in late winter may reflect increasing efforts by unpaired males to obtain mates. It would also be advantageous for a paired male to increase his efforts to defend his mate against the attentions of other males, since his chance of re-pairing, should he lose her, may well decline in late winter as established pairs disperse to their territories. A female defended by a male may also be able to feed with less disturbance from other shelducks. Ashcroft (1976) found in the eider *Somateria mollissima* that paired females have a significantly higher feeding rate than unpaired ones. She suggested that the paired female's enhanced ability to accumulate fat reserves for breeding is an important selection pressure for winter pair formation.

Consequences of dominance for the population

The hypothesis put forward by Jenkins *et al.* (1975), that subordinate individuals would be excluded from the best feeding areas and eventually from the breeding population, was only weakly supported by the Ythan studies of dominance. Subordinate males were as successful as dominants at obtaining a breeding territory and did not tend to occupy particular parts of the area (e.g. freshwater pools away from the main estuarine concentration) as the hypothesis had predicted.

There is a possibility, however, that the more subordinate males were deterred from feeding at the baited observation point. It could be regarded as a good feeding site, obviously attractive to a majority of the shelducks, which could gather a large amount of energy-rich food in a short time. The unranked males occurred much less frequently at the bait than did ranked ones which had been present on the estuary for the same length of time. Since the unranked males had been seen at the bait at least once they were obviously aware of it, although of course they may not have been attracted by grain as much as others obviously were. All the birds present at the site were scared off by the arrival of the observer at the start of each observation period and the more subordinate birds may also have been more fearful and reluctant to return. It is however possible that the subordinate males were deterred from attending the baiting site by the aggression of the more dominant birds. The grain was scattered in a fairly restricted area of shore and the subordinates would have had to approach close to dominants in order to feed. Even if the aggression of the dominants was associated primarily with defence of a mate rather than defence of the food or feeding site itself, the consequence for the subordinate would be the same, i.e. increased harassment which might well discourage further attendance.

Since shelduck numbers on the Ythan continued to increase steadily throughout the winter flock period, until after most of the pairs had taken up territories, aggression and dominance in the flock certainly did not set a ceiling to flock size and seems unlikely to have influenced the number of birds present. However, in other areas such as Aberlady, where numbers reach the breeding population level early in the winter, it is possible that such behaviour does regulate winter population size, although Evans & Pienkowski (1982) considered this unlikely. They suggested that the stability in numbers of shelduck at Aberlady in winter is caused by the tendency for both wintering and resident birds to arrive and depart on the same dates each year, and they present data on marked birds to support this hypothesis. On the Ythan, regulation of the numbers of non-breeders present later in the season is more likely. I will discuss this possibility more fully in chapter 9.

4

Territories

An animal can be said to show territorial behaviour when it has some attachment to a site (or occasionally to a moving object) and when it is aggressive towards other animals which approach that place. The resulting territory, around the site of attachment, has been defined in a variety of ways. Pitelka (1959) and Schoener (1968) emphasised the owner's exclusive use of an area, usually with defined boundaries, whereas Davies (1978) recognised territories wherever interactions between individual animals led to their being spaced apart more than would be expected from random settlement. Between these two extremes, I prefer the simple definition of territory as 'a defended area' (Noble, 1939; Nice, 1941). This embodies the essential features of a special place, around which there is aggressive defence, without implying particular features such as exclusive use or rigid boundaries, or particular consequences such as spacing out of the individuals, which may occur in many but not necessarily in all cases.

Territorial behaviour raises some interesting questions. Since aggressive defence of an area requires the expenditure of time and effort, there should be some corresponding benefit to the owner's fitness which outweighs the cost of territoriality (Davies, 1978). The spacing effect of territorial aggression would be expected to influence the dispersion pattern and density of populations. These possibilities are of particular interest in the shelduck since territorial defence, particularly by both members of the pair, is not common among ducks.

4.1 Shelduck territories

Towards the end of winter, from February to April, shelduck pairs detach themselves from the winter flock and scatter widely over muddy shores, or freshwater pools and creeks near the coast. The pairs then keep much further from others than they had done in the flock (see section 4.5) and spend most of their time alone. Marked individuals can be seen at the same spot day after day and usually threaten or attack other shelducks which come close. However, this behaviour, although clearly territorial, varies both in time and in the area defended, so that it is not always easy to define the resulting territory.

Young (1970a) followed Pitelka's (1959) definition of a territory as an exclusive area and used the term 'to refer to any area which is occupied more or less continuously by a pair of shelducks, to the exclusion of all other shelducks, regardless of whether active defence is witnessed or not.' But the exclusiveness of shelduck territories is not consistent. Jenkins *et al.* (1975) described pairs which sometimes tolerated intruders on areas which at other times were fiercely defended. They also found that some pairs moved between two sites, with other shelducks using the temporarily unoccupied one in their absence. Williams (1973) showed considerable overlap between the feeding ranges of adjacent pairs, even when both were present.

I will discuss these findings in more detail in section 4.6 but they already create difficulties for the ideas of exclusiveness and defended area, unless that area is regarded as ill defined. In shelducks, and probably in most other species, it is best to separate clearly the ideas of 'territorial' and 'territory'. Shelducks are obviously territorial in showing aggression which is localised around a site (or sites), but the resulting 'territory' is not a clearly defined area with sharp boundaries. It should be regarded as a region where other shelducks are likely to be attacked, that likelihood generally decreasing outward from the site of attachment (Patterson, 1980) and usually fluctuating in time.

4.2 The ages of territorial shelducks

Yearling shelducks can be distinguished from older ones by their plumage (section 2.2) but the age distribution among adults must

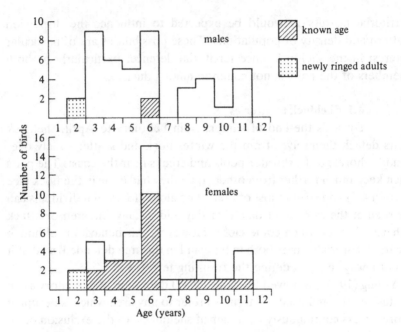

Figure 4.1 Age distribution of ringed territorial shelducks on the Ythan in 1979. The unshaded histograms show minimum age only (birds ringed as adults). From Patterson *et al.* (in press *b*).

be determined by the ages of any ringed birds. Biases may be introduced by variations in the numbers trapped and ringed in different years. This problem is most severe in the early years of a study and estimates of age composition in the Ythan population have been based on data from 1979, after a fairly constant trapping effort in 1962–5 and 1969–79.

Territorial birds ranged in age from at least two years old (newly ringed adults) to at least 11 years, with most between three and six (figure 4.1). The truncation at 11 is probably an artifact, reflecting the gap in trapping prior to 1969, but the peak of six-year-olds, especially of known-age females, followed from an unusually large production and survival of ducklings in 1973. The other fluctuations may, however, have been caused by variations in trapping effort in different years and should not be taken to show the true relative proportions of different age groups.

The male and female of each pair were often similar in age; in 36 per cent of 50 pairs in 1979 the birds were the same minimum age and in 56 per cent the two were within one year of age. Although young birds may pair together in the non-territorial flock, the main reason may well be that unringed paired adults are often trapped in the same year and so are assigned the same minimum age.

4.3 Territorial behaviour

Most territorial defence is by calling or posturing, with only occasional attacks and rare fights. The first response of a territorial male shelduck to a distant intruder or approaching neighbour is the Alert posture and obvious attentiveness. Between long-established neighbours this may be enough to induce retreat; I have seen feeding males gradually slow down, turn and move away after their neighbour adopted the Alert. If the other bird continues to advance, the resident may begin to whistle and Head-throw. The female usually Incites and joins the male and the pair may advance to meet the intruder. If it still does not retreat, the resident may attack in the Head-down, often flying since the other bird may still be at some distance. Between new neighbours there may be vigorous fighting as the attacked bird retaliates, and the two females may become involved, with male usually fighting male and female fighting female. Young (1970a) considered that only like sexes fought, but I have seen occasional attacks by territorial males on females, although this is never as common as in the mallard *Anas platyrhynchos* and blue-winged teal *Anas discors* where it is apparently the normal form of defence (Dzubin, 1955).

4.4 Establishment of territories

Throughout the winter, and commonly from early February, individual shelducks, or more usually pairs, can be seen on their own away from the flock. Sometimes the first sighting of a ringed Ythan bird in a given season was made on the site of the previous year's territory, although the bird then joined the flock for several weeks (Williams, 1973). For some pairs the change in behaviour means a change of habitat from salt to fresh water with a consequent change of diet (Hori, 1969). The period of occasional visits to the eventual territory site varies greatly in length between pairs. The average interval between first sighting on the territory and the start of continuous occupation was 25 days on the Ythan in 1971 (15 pairs, range 9–57 days) and 36 days in 1972 (9 pairs, range 13–60 days). There was no tendency for the earliest visitors to be earliest in permanent occupation (Young, 1970a; Williams, 1973).

Although visits to the territory site generally increase in frequency, each pair has a fairly sudden transition, over a few days, from being seen

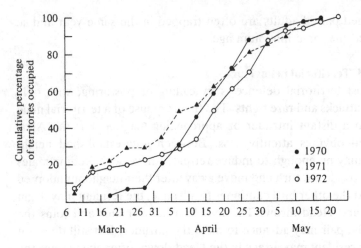

Figure 4.2 Dates of territory occupation, 1970–2. From Williams (1973).

mainly in the flock to being seen almost exclusively at the territory site (apart from visits to the separate nesting habitat, dealt with in the next chapter). Once on site, the pair spend much of their time there. Buxton (1975) found that for most pairs 97.0 per cent of sightings on the Ythan estuary (370, of 19 pairs) were on the territory, although a few pairs sometimes fed in streams or saltmarsh away from their site. Some territorial birds were also seen feeding with the non-territorial flock, especially when additional food was supplied around traps.

The quick transition to territorial occupation allows its timing to be determined for each pair as the first date of a subsequent continuous series of sightings at the territory. Almost ten per cent of pairs in 1971 and 1972 were already on their territories by early March, and occupation proceeded rapidly through April till all pairs were settled by mid-May (figure 4.2). An apparently later settlement in 1970 was probably due to less frequent observation early in that season.

The date when a pair occupies their territory depends on when the pair bond was established or re-established. Young (1970a) described two males, first seen on 18 January and 28 February 1964, which remained in the flock until 29 March when their mates were first seen on the estuary. Williams (1973) found that pairs seen together earliest were significantly earlier in taking up their territory, although there was considerable variation in the interval between pairing and settlement, both within and between years. Since males on average arrive before females (section 3.1), the taking up of the territory must usually depend on the arrival date of the female of the pair; male shelducks have not been seen to defend territories by themselves, as happens in many other species.

Other evidence, from changes of mate or of territory site between years, supports the idea of a strong female influence on territorial occupation in shelducks. A majority of surviving pairs re-form after migration each year. Young (1964*a*) found that 70.4 per cent of 71 marked birds had the same mate the following year, and 36 per cent of 25 birds retained the same mate for a third year. Williams (1973) showed that 80.5 per cent of 41 pairs remained together from year to year. Similarly, most shelducks return to the same territory site each year. In a sample of 82 which were territorial in two consecutive years, Young (1970*a*) found that 80.4 per cent held the same site. Williams (1973) found that site tenacity was greater in the female, with 80.6 per cent of 57 returning to the same site, whereas only 55 per cent of 56 males did so. The difference between the sexes was clearly linked with changes of mate; most females with a new mate still returned to their previous territory whereas males with a new mate tended to go to a new site. In all of eight cases where both members of a previous pair were present, but with new mates, the females retained their former territory while the males moved.

Deaths and experimental removal of birds also suggest that females, rather than males, 'own' the territories. Young (1970*a*) removed two males and found that their females remained on territory and obtained new mates in less than a week, whereas when two females were removed their mates abandoned the territories and rejoined the flock. The same happened after the death or disappearance of three other females and eight other males. Williams (1973) repeated these tests with similar results (table 4.5).

The shelduck thus seems to differ from most other territorial species, in which the male first takes up and defends a territory and subsequently attracts a mate or is rejoined by his previous one. The female shelduck's presence is essential to the taking up of the territory, and she apparently decides the site.

4.5 Dispersion
Dispersion of individual birds
The spatial distribution of territorial shelducks can be based either on the positions of the birds themselves or on the centres of their

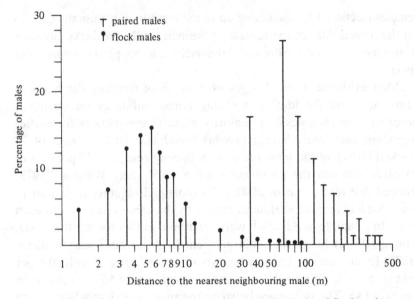

Figure 4.3 Nearest-neighbour distances of paired and flock males.
Sample sizes are 90 (paired) and 465 (flock) and are based on a number
of days' observations. From Patterson *et al.* (in press *b*).

occupied areas. The spacing of birds on the Ythan was measured by
plotting the positions of all the males on outline maps of the estuary and
subsequently measuring the distance between them. If birds were closer
than 50 m, the distance was estimated directly in the field. The resulting
nearest-neighbour distances had a bimodal distribution with a peak at
5 m and another at 50–75 m, with few males at intermediate distances
(20–30 m), from the nearest other male (figure 4.3). Most of the more
spaced-out males (over 30 m from their nearest neighbour) were in
positions where marked individuals were known to be territorial, and so
represent the degree of spacing between adjacent territorial males.
Some marked pairs, however, (14 per cent of 154 in 1978 and 1979)
although usually seen spaced apart from others, were seen infrequently
and at different locations on the estuary. Some of these 'spaced pairs'
were young adults and so may have been in the process of finding a
territory site and some may have been territorial on a freshwater pool
(where rings were difficult to identify). Such pairs were known to occur
occasionally on the estuary (Williams, 1973).

The males which occurred close together (under 20 m) were found in
a number of flocks in the wider parts of the estuary. Ringed birds in this
category were almost always seen in the close company of other birds,

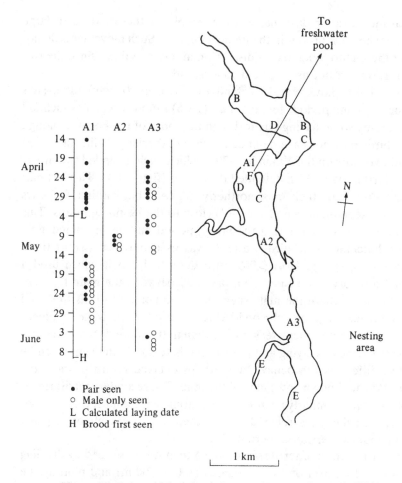

Figure 4.4 Pairs which defended more than one site on the Ythan.
Each pair is represented by one letter. The chart shows dates when pair
A were seen on their three sites. Re-drawn from Williams (1973).

forming the non-territorial flock which remained after the majority of
pairs dispersed to their territories. This flock will be discussed in detail
in chapter 9.

Dispersion of territory sites

The distribution of territories is made complex by a minority of
pairs which defend two sites. Jenkins *et al.* (1975) described three
marked birds at Aberlady which changed site during the territorial
period, and both Hori (1969) and Williams (1973) cited cases of pairs

which moved after disturbance from people or the arrival of a large flock or shelduck broods in the immediate area. Such movements do not affect the picture obtained of dispersion at any one time, since the two territories were not usually held concurrently.

Some pairs, however, do defend two or even (rarely) three sites during the same period. Jenkins *et al.* (1975) found two pairs which fed at Aberlady while having territories on the shore of the Forth, although these birds may not have defended areas at Aberlady. Williams (1973) found five pairs on the Ythan in 1970–3 which each defended two areas, and a further pair which defended three sites (figure 4.4). This last pair settled initially on their most northerly site (A1) but within a few days were also seen on a second site (A3) close to the nesting grounds. This second site was used during the laying period and again immediately before hatching. The third site (A2) was used only for a short time, again during laying. Hori (1969) also described a pair which used a second territory, on a small pool part-way between the main territory and the nest, where the male spent about 80 per cent of his time and where the female joined him on leaving the nest to feed. Male shelducks may start to wait in such areas, rather than in the more distant territory, while the female is laying and prior to the hatching of the eggs when he detects differences in behaviour (perhaps delayed return to the territory). It is not known how many of the other five secondary territories in figure 4.4 were similarly nearer the nest. Some were almost certainly not closer, and the pairs seemed to use their two areas almost equally throughout the territorial period.

The distribution of territories on the Ythan was described by dividing the area into a grid of 1.0 ha squares (100 × 100 m) and plotting the positions of the ringed birds during daily observations of the estuary (Young, 1970*a*; Williams, 1973). The territory site was taken as the square in which the pair were most often seen; even the minority of pairs with separate sites had one square which was used more than the rest.

Territories were established all over the muddy areas of the estuary avoiding only the musselbeds and sand or shingle areas (figure 4.5). Most sites had access to a water source such as the main river channel, a tributary stream or a drainage channel from a major mudflat. Most also included a stretch of shoreline above the high-water mark, used for roosting at high tide, although at exceptionally high spring tides many of the pairs roosted communally on an island in the middle of the estuary (Young, 1970*a*).

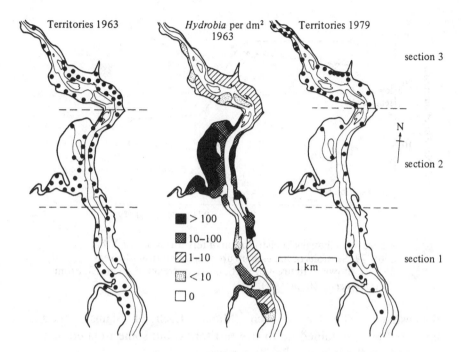

Figure 4.5 Distribution of territories on the Ythan in 1963 and 1979 and the density of *Hydrobia* in 1963. Maps for 1963 re-drawn from Young (1964*a*).

With one or two exceptions, the territories were established on areas which contained populations of *Hydrobia* (figure 4.5), the main exceptions being in the upper parts of the estuary where *Hydrobia* became scarce but was replaced by *Nereis*. Not all areas with *Hydrobia* had shelduck territories, and the density of territories did not increase where *Hydrobia* was densest. In particular, a wide mudflat in the mid-estuary, which contained the highest density of *Hydrobia*, had territories only around its edge and on its main drainage channel. The flat central area was covered only by the spring tides during the territorial period and the invertebrates, although abundant, were probably not available in the firm mud (Young, 1970*a*).

Territories on Sheppey were established along freshwater channels in low-lying grazing marshes (Hori, 1964*a*, 1969). They included the width of the water channel (about 6 m) and both banks, where birds sometimes occurred about 12 m from the water's edge. The average density was one pair per 50 m of channel (Hori, 1964*a*). The freshwater territories near the Ythan were on small lochs and pools within 5 km of

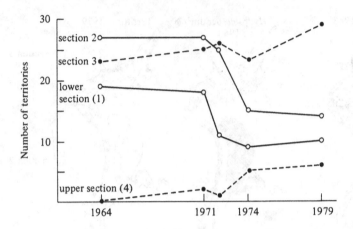

Figure 4.6 Changes in distribution of territories on the Ythan, 1964–79. Sections of the estuary are numbered from the seaward end and are shown in figure 4.5. Section 4 is upstream of section 3. From Patterson *et al.* (in press *b*).

the estuary. The largest loch (Miekle Loch of Slains), about 0.9 × 0.7 km, contained up to five territories, but some isolated pairs occupied small pools only 20–30 m across.

Changes in dispersion between years

The distribution of territories on the Ythan changed over the years, with a decrease in number in the lower part of the estuary and an increase in the upper reaches (figures 4.5, 4.6). The change occurred mainly between 1971 and 1974 and was accompanied by an increase in the number of birds which occupied freshwater pools, from a few in 1962–4 (Young, 1964*a*) to 16 in 1971–4 and a peak of 44 in 1977 (Patterson, Makepeace & Williams, in press *b*). The cause of this shift is not clear, although both disturbance by people and the growth of the green alga *Enteromorpha* may have increased on the lower mudflats in the later years. Tubbs (1977) suggested that *Enteromorpha* may reduce the suitability of mudflats for shelduck.

4.6 Territory or range size

As discussed in the introduction to this chapter, territorial defence by shelducks tends to be intermittent and inconsistent, so that the area being defended is very difficult to measure, and in any case is seldom exclusive to one pair. Instead of measuring territory size, Williams (1973) measured the range used on the Ythan by pairs in the course of their daily activities (excluding visits to the nesting area, which are considered separately in the next chapter). Burt's (1940) exclusive boundary strip method was applied to data on the number of grid squares in which each pair was seen. Most measurements used the 1.0 ha grid on the whole estuary, but since such a large grid size relative to range size tends to overestimate the area used (Stickel, 1954) more detailed data were collected in a small area with a 40 × 40 m grid (0.16 ha). This small study area was watched for whole days from dawn to dusk by two or more observers, and the position of each pair was plotted every ten minutes. The range size used in one day and changes from day to day could be estimated, whereas the data from the whole estuary, based mainly on one observation per day, gave the total area used by the pair throughout the whole territorial period.

In all such estimates of range size there is a problem of knowing when the whole range has been observed. This can be overcome by plotting an area-observation curve, the cumulative increase in the area measured with increasing number of observations. Such curves should flatten off as later observations begin to occur increasingly within previously recorded boundaries. In practice, within a limited observation period, some slow increase continues with the last observations and an arbitrary criterion is normally used to determine whether the amount of observation has been adequate. Odum & Kuenzler (1955) considered that range size was adequately measured when each new observation increased the measured area by one per cent or less. Williams (1973) used a criterion of a five per cent or smaller increase in area for each 20 per cent increase in the number of observations. He found that on his small study area with its 0.16 ha grid, 60–100 observations usually gave an adequate estimate of the day range (figure 4.7).

Using this method Williams found a mean day range of 1.66 ha, with individual ranges varying from 0.96 to 2.24 ha (table 4.1). Although each pair had a similar range size on subsequent observation days, the position of the range changed slightly. The centres of activity (Hayne, 1949) calculated for each pair for each day showed that some pairs moved over 80 m in the course of 3–4 weeks (figure 4.8). Such shifts in location mean that the size of the range used throughout the whole territorial period (estimated by combining all the observations) will exceed the size of area used in any one day. Williams (1973) found that for the eight pairs in his study area, the whole season's range exceeded the maximum range for one day by 59 per cent (table 4.2) Pairs were variable in how much they shifted, with whole season's ranges varying

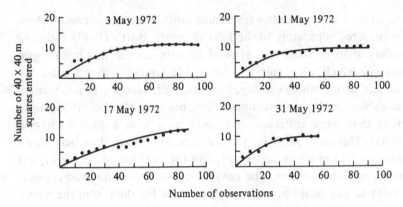

Figure 4.7 Area–observation curves for the same pair on different days in May 1972. Modified from Williams (1973).

Table 4.1. *Sizes of day ranges in 1972, measured on a* 40 × 40 m *grid. Each line refers to one pair. Modified from Williams (1973)*

Pair					
	3 May	11 May	17 May	31 May	mean
1		1.28	1.84		1.56
2	1.60	1.28		1.52	1.47
3		2.08	1.60	2.08	1.92
4		2.00	2.00		2.00
5	0.96		1.12		1.04
6		1.04			1.04
7	2.00	1.92	1.76		1.89
8		2.24	0.96	2.24	1.81
mean	1.52	1.69	1.55	1.95	1.66

from 14 per cent to 100 per cent larger than their largest day range. He considered that to avoid such shifts of location, but yet make certain of adequate measurement of the range, observations every ten minutes from dawn to dusk on two consecutive days would be necessary to estimate shelduck range size.

For the majority of pairs on the Ythan, range size was estimated from one observation per day using the 1.0 ha grid. The effect of this could be checked for the eight pairs on the small study area by estimating their season's range sizes by both grids (Williams, 1973). As expected, range size was larger, on average by 67 per cent, using the larger grid square. Since estimates based on smaller grid squares are likely to be more

Table 4.2. *Comparison of range sizes (ha) estimated from one day's data and from the whole season's data. The pairs are those in table 4.1. From Williams (1973)*

Pair	Maximum day range	Whole season's range	Percentage difference
1	1.84	2.72	48
2	1.60	2.24	40
3	2.08	3.20	54
4	2.00	3.60	80
5	1.12	2.24	100
6	1.04	1.76	69
7	2.00	3.84	92
8	2.24	2.56	14
Mean	1.74	2.77	59

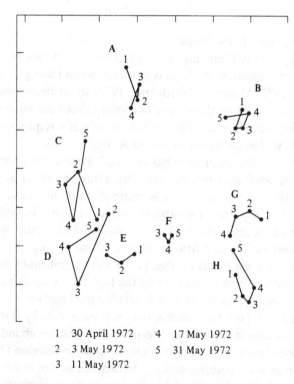

1 30 April 1972	4 17 May 1972
2 3 May 1972	5 31 May 1972
3 11 May 1972	

Figure 4.8 Shifts in the centres of activity of neighbouring pairs. The grid shown is of 40 × 40 m squares. Modified from Williams (1973).

Table 4.3. *Mean size of whole season's ranges of pairs on the Ythan estuary. From Williams (1973)*

Year	Pairs	Mean number of sightings per pair	Range size (ha) Mean	Range size (ha) Standard error	Range
1970	29	26.7	3.2	1.34	0.67–6.67
1971	39	70.0	4.0	1.28	1.67–7.00
1972	36	21.4	2.2	0.90	1.00–5.00

accurate, Williams correspondingly reduced his measurements made on the 1.0 ha grid. This gave an estimate of 2.2–4.0 ha for the mean whole-season range of the Ythan shelducks (table 4.3). The value of 4.0 ha for 1971 is likely to be the best estimate, since the frequency of visits to the estuary was increased to four per day in that season, giving a roughly three-fold increase in the mean number of observations per pair.

Territory size and food supply

Territory size is frequently correlated with food supply, such that territories are smaller where food is more abundant (Stenger, 1958; Cody & Cody, 1972; Slaney & Northcote, 1972) or of better quality (Moss, 1969). Each territory then tends to contain about the same total quantity of food, sufficient for the individual animal's requirements, e.g. in sunbirds *Nectarinia reichenowi* (Gill & Wolf, 1975).

The abundance of the main invertebrate food species can be measured on shelduck territories by core sampling (Buxton, 1975) but the availability of this food to shelducks is more difficult to assess. The shore is covered by deep water for variable periods at high tide, while at low water many of the invertebrates burrow deeply into the mud, which itself may become too dry and firm to allow efficient sieving.

Buxton found that territories on the Ythan were significantly larger on areas which were covered for longer by the high tide, suggesting that territory size was related to the area of accessible mud and the time for which it was exposed. However, preliminary measurements by Bateson (1968) showed no correlation between territory size and the abundance of the prey species in the mud. A more detailed study by Buxton (1975) showed a tendency for territories actually to increase in size in areas of higher food abundance. In one of his two study areas territory size increased significantly with increased density of *Hydrobia* (figure 4.9).

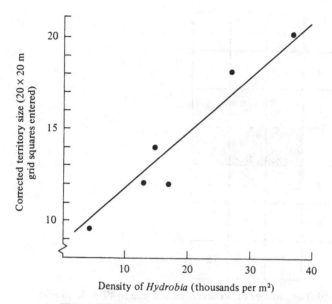

Figure 4.9 Territory size in relation to *Hydrobia* density. The line is a calculated regression ($y = 8.50 + 0.003x$, $r = 0.942$, $p < 0.01$). From Buxton (1975).

With *Corophium* and *Nereis* and with all three species in his other area there was a similar, though non-significant, trend or no correlation. Measurements of the abundance of prey do not of course allow for variations in its availability to the birds and the proportion of the birds' time that they spend feeding (which may be related to the rate at which food could be obtained, i.e. to both availability and abundance) was not correlated with territory size.

Shelduck territory size thus clearly does not decrease with increasing food density, as might be expected. The birds, however, seem only to settle where suitable food is present, presumably above a minimum threshold abundance. Buxton's finding that territories were actually larger where food was denser is particularly puzzling and suggests some other factor, positively correlated with both territory size and food, affecting territorial behaviour. It is possible, for example, that large open areas of mudflat lead to larger territories and may also coincidentally have higher invertebrate densities.

Range usage and overlap

Shelducks do not use all parts of their range equally but are seen in some parts more than in others. Williams (1973) counted the number

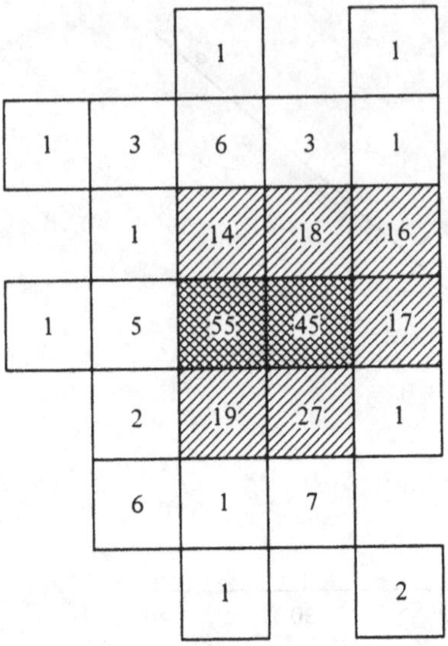

Figure 4.10 The number of occurrences of a pair in 40 × 40 m grid squares. The most frequently used squares are shaded. Modified from Williams (1973).

of occurrences by each pair in each 40 × 40 m grid square of their range and found that most pairs confined their activities to only a few of the squares (figure 4.10). The distribution of sightings over the squares was highly non-random, although in the example shown the usage of the two most commonly used squares did not differ significantly from equality. Such randomly and highly used squares were identified by eliminating progressively the less frequently utilised ones (first those with five or fewer sightings, then those with ten or fewer, etc.). A test of randomness (χ^2) was applied at each stage, until the distribution of sightings did not differ significantly from equal use. This determines those most commonly used squares in which the frequency of occurrence was approximately the same. Occasionally only one 40 × 40 m square was used significantly more than all the others, and sometimes the squares most often used were not contiguous.

Overlap of range between adjacent pairs of shelducks is usually considerable, in spite of Young's (1970a) reference to 'mutually exclusive feeding territories'. The eight pairs in Williams's (1973) study area had a mean overlap of 91 per cent of their season's ranges by other pairs (range 76.5–100 per cent). The total area of mudflat on the Ythan was around 184 ha (A. Anderson, pers. comm.) of which Williams estimated about 130 ha were occupied by territorial shelducks. Thus in

1971, when 68 pairs were territorial on the estuary, only about 2 ha of mudflat were available for each pair. Since mean range size was 4.0 ha in 1971 there must have been a considerable degree of range overlap between pairs, although this need not have occurred when adjacent territories were both occupied simultaneously; pairs readily expanded their ranges into areas used by their neighbours when the latter were temporarily absent. Williams, although unable to estimate the precise degree of overlap when adjacent pairs were present simultaneously, considered it to be around 25 per cent of the total range. Almost all of this overlap was in the least-used squares of each pair's range; the most frequently used areas were rarely overlapped by other birds and then always when the owners were absent. The greater degree of overlap between whole-season's ranges will of course include the successive use of the same area by different pairs through shifts in territory location.

4.7 Use of the territory for feeding
Attendance
Attendance at the territory site varies through the territorial period, which can be divided into the pre-laying, laying and incubation phases. These are determined by estimating the laying date from the date of appearance of the brood of each successful pair and from the pattern of arrival and departure of the male and female. During the pre-laying period the male and female are often absent together for some hours just after dawn while they prospect for nest sites (see section 5.1). Laying is the most difficult phase to identify by observation since the female is absent for only one short period each day and the male may stay away from the territory while she is at the nest (Buxton, 1975). During incubation the male is alone on the territory for most of the day, with the female making a number of short visits.

Attendance has been measured on the Ythan by two methods. Williams (1973) made spot checks on 25 marked pairs around the whole estuary in 1970 and 1971 and recorded whether the birds were present at each check (which could be at any time or stage of tide). Both he and Buxton (1975), with assistance from myself, made dawn-to-dusk observations in 1970–2 on groups of up to eight adjacent territories at one time, and recorded the presence and absence of the birds every ten minutes. Such continuous observations gave much more accurate estimates, with actual arrivals and departures often being timed exactly, and were spread evenly through the daylight period.

During the pre-laying period both male and female were present on

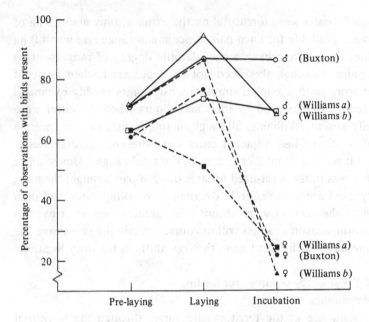

Figure 4.11 Attendance at the territory site over the territorial period. Square symbols show data based on spot observations over the whole estuary (Williams, *a*). The other data are based on dawn-to-dusk observation. Drawn from data tabulated in Buxton (1975) and Williams (1973).

the territory for 60–70 per cent of the day (figure 4.11). The birds were already present at dawn and were still there when darkness fell, so it was likely that they had stayed throughout the night. They were thus present for around 80 per cent of the whole 24 hours (Williams, 1973). All but one of the estimates showed higher attendance during the laying period, and each estimate showed the female to be absent more than the male who was present on the territory for much of the day. A low attendance during laying was shown only by females in Williams' estimate from spot checks on the whole Ythan (figure 4.11). This may have resulted from the inclusion in this sample of long absences at the nest which were recorded for some females towards the end of egg-laying (see section 6.2). At the start of incubation, attendance at the territory by the female declined dramatically to 14–24 per cent of her day, concentrated into two or three visits (see section 6.4). Males also showed some decrease in attendance from the high level during laying, to 68–86 per cent of the day (figure 4.11). These estimates of male attendance are averages for the whole incubation period; Williams (1973) found that males generally

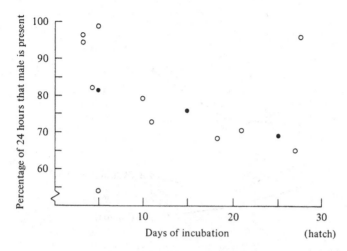

Figure 4.12 Proportion of time spent on the territory by males through the incubation period. Open symbols show data from dawn-to-dusk observations; closed symbols are mean values for each 10 day period. From Williams (1973).

spent less of their time on territory as incubation progressed (figure 4.12) (although of the 11 males observed, one showed low attendance early in incubation and one showed almost constant presence at the end). The drop shown overall was probably caused by males spending increasing amounts of time near the nest or on a secondary territory site on the way to the nesting area. Finally, pairs which failed to hatch their eggs adopted a pattern of attendance similar to that during the pre-laying period. Williams found that such males and females were present for 62.6 and 62.2 per cent of their time respectively (total time 159 h). Successful pairs usually abandoned the territory when the young hatched and the brood was led to another area (see section 7.4).

Feeding

The proportion of time spent feeding by shelducks on their territories was measured on the Ythan by Buxton (1975) and Williams (1973) from the same dawn-to-dusk observations used to estimate attendance. Every ten minutes the behaviour of each bird present was recorded, particularly noting whether or not it was feeding. At all stages of the territorial period, females fed for more of their time than did males (figure 4.13). This was particularly marked during incubation, when females fed for 80–90 per cent of the short time they spent on their

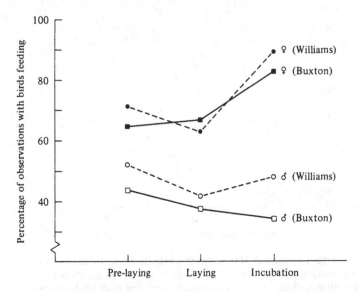

Figure 4.13 Proportion of time spent feeding while on territory during the territorial period. Drawn from data tabulated in Buxton (1975) and Williams (1973).

territories. Apart from feeding they usually spent a short time bathing and preening immediately before returning to the nest. Williams found that females which failed to hatch their eggs spent only 57.3 per cent of their time feeding after abandoning the nest. This was markedly lower than in the pre-laying period (71.2 per cent) and only just above the level shown by their males (51.9 per cent), strongly suggesting that such females were not attempting to build up reserves for a repeat clutch.

The above measurements of the proportion of time spent feeding while present on the territory (figure 4.13) must of course be taken together with estimates of the proportion of the day spent on the area (figure 4.11) to assess the total number of hours for which the birds feed each day. For example, incubating females feed for much of their short periods on territory whereas their mates feed for less of their time over more of the day. Having assumed that shelducks (except incubating females) feed at night as much as by day (as they do in winter), Buxton (1975) estimated the total number of hours of feeding for males and females in a 24-hour period throughout the season. Females were much more variable than males in the number of hours spent feeding per day and, during the pre-laying and laying period, fed for 3–4 hours more than their mates (table 4.4). The difference was particularly marked

Table 4.4. *Mean number of hours spent feeding per 24 hours during the breeding season. Modified from Buxton (1975)*

Phase of season	Hours spent feeding per 24 hours		
	Male	Female	Difference
Pre-laying	6.33	9.35	3.02
Laying	7.74	12.18	4.44
Incubation	6.96	3.03	3.93

during laying, when females fed for 12 hours out of the 24, and presumably reflects the energy required for egg production, and possibly also for a build-up of fat reserves prior to incubation. It is not clear why males should show a slight increase at the same time, unless they were responding to their mates' high level of feeding. During incubation, females fed for only 3 hours per day, only one quarter of the time spent feeding before the start of laying. The amount of food obtained may not have been reduced to the same extent, however, since females appeared to feed particularly voraciously during feeding breaks in the incubation period.

Factors correlated with the proportion of time spent feeding

During the pre-laying and laying phases of the season, Buxton (1975) found that the proportion of time spent feeding by the female of a pair was significantly correlated with that of her mate, suggesting that male and female feeding is affected by the same factors. Females spent significantly less of their time in feeding if their territories had a higher density and biomass of *Hydrobia*. There was no correlation with the density or biomass of *Corophium* or *Nereis*, confirming again that *Hydrobia* was the most important prey species.

Females with a higher water content in the mud of their territories also spent significantly less of their time feeding (Buxton, 1975), presumably because such 'wetter' territories were found to have a significantly higher density and biomass of *Hydrobia* and a significantly higher biomass of *Nereis*. The food may also have been more accessible in wetter mud. Thus, although territory size was not related to food abundance, the amount of food present benefited the birds by reducing the amount of time they had to spend feeding during the pre-laying and laying period.

Incubating females showed no change in the proportion of time spent feeding with change in the density or biomass of the prey, presumably because at this stage the birds feed for almost all of their short stay on the territory. The mean length of stay on the territory by incubating females should be shorter on territories with higher food abundance, but this was not tested by Buxton (1975).

4.8 Benefits and costs of being territorial
Benefits

In shelducks, we can eliminate some of the benefits of territoriality which accrue to other territorial species. The territory does not ensure that the pair have a nest site, since it is purely a feeding area remote from the nesting habitat where the birds are not territorial (see section 5.4). Similarly, pair formation takes place in the flocks and single males do not take up territories and use them for advertisement. Hori (1969) has suggested that the territory might reduce the chance of interference with copulation, common in many species of duck (Hochbaum, 1959). However, I have not been able to find any direct evidence of such interference in shelducks. Hori argues from the spread of excited diving in flocks leading to some mountings, whereas observations by Young (1970a) and myself suggest that territorial pairs always copulate in water, which is often at the edge of the territory, close to neighbours on the other bank and in a relatively undefended area. I have also seen pairs copulating in freshwater pools in the nesting area, so that a territory is not essential for this. Finally the territory does not reserve a feeding area for subsequent use by the young, since these are almost always taken to a new area (Hori, 1964a, 1969; Young, 1964a; Williams, 1974).

Territorial behaviour in shelducks is closely associated with nesting. Only territorial pairs appear to lay, and non-territorial birds have never been recorded with a brood (Young, 1964a; Williams, 1973). The territory is taken up when the birds begin to prospect for a nest site (see section 5.1) and defence continues through laying and incubation till the eggs hatch. This suggests some benefit of a territory to nesting pairs.

The territory site could act as a meeting place for the pair if they become separated (Hori, 1969), particularly during laying and incubation since they are rarely apart at any other time (Williams, 1973). Hori described how mallard females in early incubation met their mate at an habitual loafing site before going to feed. However, shelduck males sometimes stay near the nest or on a secondary site while the female is sitting, and are only on the territory for 70–85 per cent of the time during incubation. The female calls loudly and regularly while flying from the nest to feed and I have often seen sleeping males wake up and fly to meet their returning mate. I have also seen females, arriving at the territory to find the male absent, flying up and down the shore calling loudly. The mates of such females usually re-appeared quickly, so there seems little difficulty in meeting without having a territory, as long as the male stays within earshot.

A much more likely benefit of a territory is that it provides an exclusive, undisturbed feeding area for the breeding female during the pre-laying, laying and incubation periods. Territorial females are able to spend almost all of their time feeding when off the nest during incubation. Williams (1973) found that they spent only 0.4–1.4 per cent of their time in aggressive behaviour on the territory during the nesting period, whereas non-territorial flock birds were involved in frequent interactions (see section 9.6), which presumably interrupted feeding. Ashcroft (1976) found that eider females protected by their mate had a higher feeding rate than had single females. Pienkowski & Evans (in press *b*) also considered that the main benefit of a shelduck territory is to provide an exclusive, undisturbed feeding area for the female. They found that the feeding rate was halved (from 44.6 ± 4.2 to 22.6 ± 6.9 pecks per min) when a territorial pair were involved in an aggressive interaction, and that feeding was also disrupted when pairs were kept off their territories.

The importance of undisturbed feeding depends on whether the female's time for feeding is critically short, particularly during incubation. Ingestion rates have not been measured in shelducks, since their filtering technique makes observation impossible, but comparison of the time spent feeding at different stages of the season gives some indication of the amount of feeding time necessary to obtain the bird's food requirements. Buxton (1975) found that females in mid-March in one of his study areas fed for only 14.85 per cent of the time during a full tidal cycle, and that feeding at night seemed to be similar to that during the day. These birds were presumably able to collect their food requirement in this time, i.e. 3.56 h. However, he mentions that some birds at that time were seen eating grain in fields, so that their estuarine feeding might have been reduced. Females in early March fed for 35.6 per cent of the time (mean of two study areas), i.e. 8.5 h per 24. This might be taken as the maximum time necessary for the female to collect her food requirement, given that temperatures would be lower, and hence energy requirement for thermoregulation higher, than in the nesting season.

Aggressive activity was rising at that time (section 3.5) and the females may have needed to gain weight for breeding, both of which would have increased food requirement. An alternative estimate of maintenance requirement is the time spent feeding by the territorial male during the nesting season. This varied from 6.33 h to 7.74 h per 24 h (table 4.4), somewhat lower than the 8.5 h estimated above. Finally, King (1973) estimated that the daily cost of egg production in Anseriformes is 52.7 per cent above normal maintenance energy requirement. Laying female shelducks fed for 12.18 h per 24 h (table 4.4) suggesting that they might satisfy maintenance requirements in 8.0 h per 24 h, which is fairly close to the estimate from feeding in March (above).

If a female shelduck does require 8.0–8.5 h of feeding time to gather her maintenance requirements, incubating females which feed for only 3.03 h per day (table 4.4) would seem not to be maintaining themselves. There may, however, be an increased rate of ingestion (judging from the hurried appearance of such females), which may partly offset the short feeding time. Their energy requirements while sitting relatively still in a very sheltered nest site may well be reduced; most estimates of the energy costs of incubation suggest that this is less than the cost of normal daily activity and sometimes little more than that of the resting bird (Mertens, 1980).

Limited data on weights of incubating females suggest that they do not lose condition. Ten females caught in May while incubating (stage unknown) weighed 950.8 ± 19.3 g while eight non-incubating territorial females caught at the same time weighed 1001.3 ± 34.9 g and 31 flock females weighed 919.7 ± 16.3 g (see figure 2.1). This suggests that females do manage to collect their maintenance food requirements in the short time they spend off the nest.

It seems likely that incubating female shelducks must balance two conflicting demands: to feed for long enough to maintain condition so as to be able to care for the young after hatching, and to stay away from the eggs for as short a time as possible to reduce risks from chilling and possible interference by other females (see section 6.3). In maintaining this balance it would be advantageous to the female to feed as quickly and efficiently as possible. Having an assured, exclusive and undisturbed feeding area is, therefore, likely to ensure this.

Costs

The costs of being territorial are chiefly the time and energy used in threatening and attacking other birds and the loss of feeding time through watching for potential intruders. Most of this cost is borne by the male, since he carries out most of the attacks, females spending very little time in aggression (see above). Williams (1973) found that territorial males spent 1.8–4.7 per cent of their time in aggressive behaviour and a further 7.7–12.9 per cent in the Alert posture. When incubating females were off the nest their mates were Alert almost all of the time and most males rarely fed while their mates were there. The females, in contrast, rarely became alert, presumably benefiting from the males' constant alertness by being able to feed continuously. The time spent in aggression and Alert by males was unlikely to have a serious adverse effect, however, since they normally spent less than 50 per cent of their time feeding (figure 4.13) and Williams found that they spent 17.6–29.4 per cent resting and preening. Thus, they seem to have plenty of time for additional feeding if necessary, and indeed territorial males give the impression of feeding in a rather leisurely way, especially while their mate is incubating. The costs of being territorial thus do not seem to be very large in shelducks.

4.9 Consequences of territoriality for the population

Territorial behaviour can limit the density, and consequently often the size, of local populations through the exclusion of potential settlers, which would have been capable of establishing a territory (Patterson, 1980). This has been demonstrated by removal experiments in the red grouse (Watson & Jenkins, 1968), the crow *Corvus corone* (Charles, 1972) and many other species. If territorial behaviour changes in relation to food abundance, so as to satisfy the requirements of the territorial individuals, density will be adjusted to food supply (Patterson, 1980). In the red grouse, for example, males are more aggressive in years of poor food supply, resulting in larger territories and a lower breeding density (Watson & Moss, 1972).

In the shelduck, territorial behaviour has been suggested as a possible factor limiting breeding density; indeed Hori (1969) and Young (1970a), following Wynne-Edwards (1962), considered that the behaviour had been selected mainly for this function. Even without this interpretation, which implies selection at the population level, territorial behaviour selected for its individual benefits might still tend to limit population density.

Both Young (1970a) and Williams (1973) carried out removals of established pairs to test whether others were being excluded from settling. Young removed four pairs and found that their territories were

Table 4.5. *Experimental removal of territorial pairs. From data in Young (1964a, 1970a) and Williams (1973). Estimated percentage of pairs territorial on each date from Williams (1973)*

Category removed	Date of removal	Estimated percentage of pairs territorial	Date of replacement	Interval (days)
Pair	24.2.64	0–10	26.3.64	31
	3.4.64	20–40	8.4.64	5
	14.4.64	30–55	27.4.64	13
	29.4.64	70–90	4.5.64	5
	10.4.71	30–55	21.4.71	11
	12.5.71	97	Neighbours took area	*No replacement*
Female only	13.4.64	30–55	Male abandoned area	Pair replaced
	5.5.64	87–95	Male abandoned area	Pair replaced
	13.4.71	53	Male abandoned area	Pair replaced
	16.5.72	98	Male abandoned area	*No replacement*
Male only	28.4.64	70–90	4.5.64	6
	12.5.64	94–98	13.5.64	1
	12.4.71	53	New male replacement	
	15.5.72	98	New male replacement	

taken over fairly quickly by formerly non-territorial pairs (table 4.5). When two females were taken, their males abandoned the territories, which were then taken up by new pairs. Thus these experiments seemed to show that there were potential territorial settlers in the non-territorial flock, prevented from settling by the presence of the original territorial birds. This conclusion, however, was criticised by Williams, who pointed out that all of the removals were made before the end of April (indeed one was in February) at a time when not all of the territorial pairs had taken up their territories (figure 4.2). Some or all of the replacement pairs might have obtained territories elsewhere in any case, even if in less attractive places. Williams (1973) repeated the same test but removed some of the birds in mid-May, when all the territories for that season should have been established and any further pairs would have remained non-territorial. The April removals were all replaced as in the earlier study but those in May were not (table 4.5). (One May removal was perhaps not a fair test, in that the vacant territory was immediately taken over by neighbours and thus was not available for potential settlers. This leaves only one certain non-replacement.) Both workers found that when males were removed their mates quickly re-paired on the same territory, suggesting that there were surplus unpaired adult males in the non-territorial flock.

From these results, Williams suggested that there was no evidence of limitation of density through territorial behaviour, since vacant territories were filled only before May, when the incoming pairs might well have gone on to be territorial elsewhere. However, his argument rested heavily on one crucial May removal and more are obviously needed. In addition the experiments in 1971 and 1972 could not be exact replicates of those in 1964 since the size of the non-territorial flock, and hence the number of potential settlers, differed between the two periods. The flock in 1964 was much larger and probably contained more adults than did the flocks in 1971 and 1972 (figure 10.3). Replacement was thus much more likely in 1964. More removal experiments need to be done in May in years when the flock size is large.

There is thus no unequivocal evidence that territorial behaviour can limit population density in shelducks. Jenkins *et al.* (1975) also concluded that 'territorial behaviour was not essential for division of the stock into breeders and non-breeders', since at Aberlady the number of birds levelled off and the population divided into groups of different status in January or early February, some two months before territories were established. Pienkowski & Evans (in press *b*), however, released birds at Aberlady after most territories were taken up and found that an equivalent number left the Bay. It may be that territorial behaviour would begin to limit the number of breeding pairs only where the population had become very large in relation to the amount of suitable habitat.

Adjustment of density to food supply is unlikely to be achieved through territorial behaviour, since territory size was not found to be related to the abundance of the prey. Indeed the opposite occurred, in that territories were larger in areas of high prey density than in low density areas.

If territorial behaviour has been selected because it provides an undisturbed feeding area for the female, as I have suggested, the principal effect on the population will be a non-random spacing out of the breeding pairs so that mutual disturbance is minimised. Each pair should occupy about the same amount of space, irrespective of food abundance (above the minimum required to meet requirements). The fact that territories were larger in areas which were covered by the tide for longer periods supports the idea that a particular area of accessible shore is required by each pair. If so, the density of territorial pairs should have a similar maximum in different areas. Unfortunately, territorial density has rarely been measured (see section 8.3) so that this cannot be tested quantitatively.

4.10 Abandoning the territory

Shelduck pairs which succeed in hatching ducklings usually take them to a new brood range (see section 7.4), leaving their territory unoccupied. Pairs whose nesting attempt fails usually stay on the territory for some time before abandoning it and joining the non-territorial flock prior to the moult migration. On the Ythan, most failed breeders left their territories between mid-May and the end of June, with a peak in mid-June. The duration of stay after failure depended on the date when the nest was lost, estimated from when the male was last seen alone on the territory. Pairs which failed before 20 May stayed 16–34 days. Those failing later in May stayed 6–25 days and those failing in June stayed only 1–9 days (Williams, 1973).

5

Prospecting for nest sites

The selection of a nest site poses considerable problems for a bird; in particular the site must be safe from potential predators of the clutch, or of the incubating bird, throughout the prolonged period of laying and incubation. Some species select inaccessible sites in trees, cliffs or holes, some disperse their highly cryptic eggs far from nests of the same species, whereas others rely on grouping closely into colonies where communal defence may deter or drive away predators. It is imperative that the bird makes the correct choice, since on it may depend the only chance of producing progeny, a major component of genetic fitness. In this chapter I will describe the behaviour of shelducks prospecting for nest sites and will discuss some of the problems they encounter.

5.1 Timing of prospecting

From around mid-March, pairs of shelducks begin to visit potential nesting places (Gillham & Homes, 1950; Hori, 1964a, 1969; Young, 1970b). Each pair starts around the date when it establishes its feeding territory. In a sample of 45 Ythan pairs, 16 per cent were first seen in the sand dune nesting area just before they were seen on territory. A further 44 per cent were first seen in the dunes during their

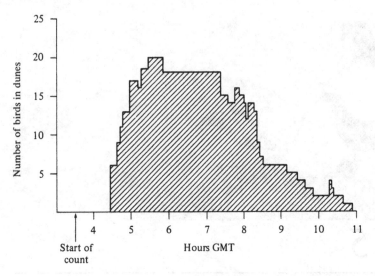

Figure 5.1 The number of shelducks in the Ythan dune study area, from counts of birds flying in and out over the west boundary, 1 June 1972. From Patterson & Makepeace (1979).

period of territorial establishment and 40 per cent just afterwards (data on territorial behaviour from Williams, 1973). Gillham & Homes (1950) and Hori (1964a, 1969) refer to initial gatherings in grazing marshes near the shore in March, before the birds extended their visits to more distant nesting areas. I have seen similar small groups on the Ythan salt-marshes and on fields between the estuary and the dunes, perhaps representing an early stage of nest-prospecting when the birds are not sufficiently motivated to enter the nesting areas proper. By early April pairs and groups of shelducks can be seen throughout the nesting habitat.

Visits to potential nest sites are made almost entirely in the morning, starting around sunrise (Hori, 1964a, 1969; Young 1970b). The diurnal pattern of attendance can be determined by counting the birds as they fly between the territorial and nesting areas. The Ythan birds moved rapidly into the dunes just after dawn and most remained at least 90 minutes before starting a gradual return to the estuary, with the last birds leaving some six hours after the first arrivals (figure 5.1). About 50 per cent of the shelducks still remained three hours after arriving and this period was used for observations of their behaviour. Pairs were seen only occasionally in the afternoon or evening in the Ythan dunes, but Hori (1969) described groups at Sheppey which remained in the nesting area all day. This difference was probably not due to disturbance by people, who rarely entered the Ythan dunes except at weekends, but may represent a real difference in behaviour between the two populations.

The synchrony of visits to the nesting habitat by most pairs in the few hours after dawn will enhance the possibility of grouping together, since the maximum number of pairs will be in the area at the same time. Dawn is a convenient timer, commonly associated with other kinds of social behaviour such as bird song. Possibly there is a selection for grouping, which I will discuss later. Other possible functional explanations of the timing of visits seem less likely; the birds do not feed at a particular time of day but at particular states of tide (section 2.5), the nest sites are unlikely to change during the day and the risks from predators are unlikely to be less in the early morning. Indeed almost all sightings of foxes *Vulpes vulpes*, hunting in the Ythan dunes in daylight, were made just after dawn and there was little disturbance from people there at any time of day. Hori (1964*a*), however, suggested that many of the shelducks on Sheppey were disturbed by farm workers later in the morning.

Seasonal variations in the number of shelducks visiting the nesting area can be measured by making standard counts during the three-hour period when most birds are present. On the Ythan, I obtained data on numbers and dispersion by walking along a standard route through the dunes, keeping out of sight of the birds as much as possible. The position of each shelduck seen could be recorded by using a map prepared from aerial photographs. The total number seen on the ground could be used only as an index of the true number of birds present, since some moved during the count and an unknown proportion could have been missed or counted twice. The error, however, was probably only small; an alternative method, of counting the birds as they flew from the estuary to the dunes, was used three times as a check in April–June 1972, while another person made the standard walking count. The flight count totals of 25, 14 and 23 (mean 20.7) were similar to those from the standard method (23, 12 and 32, mean 22.3).

The number of shelducks present in the Ythan dunes showed a distinct peak in mid-April and another in late May or early June, with very few birds present in early May (figure 5.2). This period of low numbers coincided with a peak of egg-laying, calculated by Williams (1973) from hatching dates in the same years, and presumably reflected a cessation of nest-prospecting by females which had started to lay. The

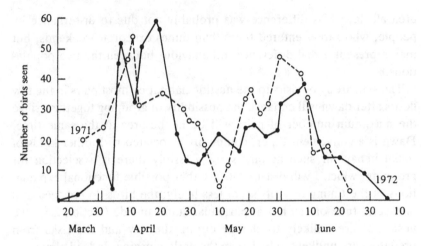

Figure 5.2 Seasonal variation in the number of shelducks seen on the ground from a standard counting route in the Ythan dunes. From Patterson & Makepeace (1979).

first peak of attendance was thus about two weeks before the peak of egg-laying, and some marked pairs were seen in the dunes for at least a month before their estimated laying date.

The age, sex and territorial status of the birds taking part in visits to the nesting area can be determined from the distinctive plumage of the yearlings, from the pronounced sexual dimorphism and from other data on individually marked birds. All the shelducks seen in the March–April peak on the Ythan were in adult plumage (at least two years old) and there were equal numbers of males and females, reflecting the preponderance of adult pairs (table 5.1). In May and June, about 8 per cent of the birds were yearlings, all of them females, but the sex ratio among

Table 5.1. *The ages and sexes of shelducks visiting the Ythan nesting area. From Patterson & Makepeace (1979)*

| | | Adult | | Yearling | |
		male	female	male	female
March–April	1971	118	118	0	0
	1972	205	195	0	0
May–June	1971	188	188	0[a]	26
	1972	130	132	0[a]	25

[a] Sign test against an equal sex ratio, $p < 0.001$.

the adults remained equal. Over the same period, Williams (1973) determined the territorial status of the marked individuals. Most of the adults attending in the March–April period were territorial and many succeeded in hatching broods, while those present in May and June included many two and three year old birds without territories. For example, in 1975, all of 12 pairs seen before mid-April, but only five of 11 first seen later, were territorial (Fisher test, $P = 0.005$). Those territorial pairs which arrived late in the dunes were very unsuccessful; in data from 1970 to 1972 and 1975, none of ten pairs which were first seen in May and June succeeded in hatching a brood, whereas 31 (43 per cent) of 72 pairs which arrived before then were successful ($\chi^2 = 5.21$, $P < 0.05$). In addition, a number of pairs which first appeared in the dunes in March and April, but which apparently lost their clutches, also occurred throughout May and June. Hori (1969) also described failed breeders which continued to visit the nesting area for the rest of the season. Young (1964*a*) did not record the arrival date of each pair in the nesting area but he made only one observation before mid-April and made 70 per cent of his sightings of pairs in May and June. He similarly found that these late-occurring pairs were unsuccessful, with only 13 per cent (of 39 pairs) hatching a brood.

Thus, there are two clearly separate phases in prospecting for nests, involving two largely separate groups of shelducks. Those visiting the Ythan dunes in March and April were territorial adult pairs and the low numbers seen in early May were presumably due to the onset of incubation among those early pairs. Some of those which failed resumed their visits, but most of the pairs visiting the area in May and June were new, non-territorial and young pairs (which possibly come into breeding condition later than older ones) and very few of these nested successfully. The marked preponderance of females among yearlings visiting the dunes suggests either that the sexes mature at different ages or, more likely, that visits to the nesting area are made primarily by females and that males visit only when they are paired to visiting females. Yearling males were seldom paired and were seen so rarely in the nesting dunes that none appeared in the sample in table 5.1.

Why should shelducks persist in visiting the nesting area so late in the season when so few of them are likely to be successful in nesting? The problem is most marked in the non-territorial pairs and unpaired yearling females, both categories which have virtually no chance of breeding. It is likely that these birds may gain familiarity with the nesting habitat and nest sites, which could aid future site selection, and

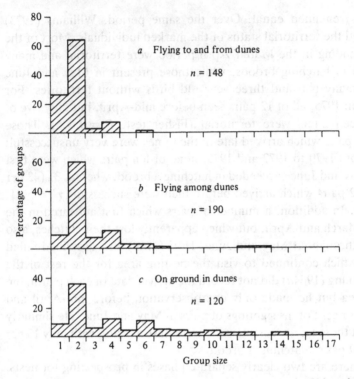

Figure 5.3 Group size among shelducks seen in the Ythan dunes.

they may be able to compare different parts of the habitat for frequency of disturbance, presence of potential predators and other factors which might affect breeding success. There may also be the possibility of parasitic laying, which I will discuss later.

5.2 Grouping

Shelducks visiting the nesting habitat are gregarious, although they are aggressively territorial while on the feeding areas at the same stage of the season. Pairs leave their territories independently of others and, although they may be joined by other pairs on their way to the nesting area, 64 per cent of groups flying to and from the nesting area were single pairs (figure 5.3). Shelducks seen flying around among the Ythan dunes in 1970–2 were also most commonly in single pairs (figure 5.3). A further 17 per cent of single birds were mainly females leaving the nest to feed and males returning to their territories after accompanying their incubating mates back to the nest, although many yearling

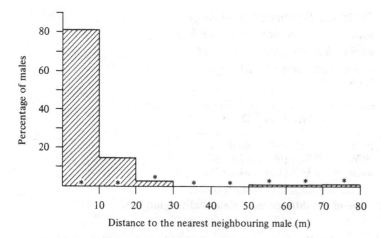

Figure 5.4 Nearest-neighbour distances between males in the Ythan dunes, 1976. The histogram shows observed distances (n = 608). The expected distribution (asterisks) is based on the mean number of males (12.15) observed in the study area (160 ha).

females also flew alone. Only 30 per cent of the groups were of more than two birds and most were under five.

Once in the nesting area, however, the pairs generally join others already there. Gillham & Homes (1950) refer to groups of 'a score or more' and to 'social groups' in the marshes, while Hori (1964*a*) described 'display gatherings', 'breeding groups' and 'miscellaneous gatherings'. The Ythan birds also often formed groups, although single pairs were still the commonest among groups seen on the ground (32 per cent, figure 5.3*c*). However, larger groups (up to 16) were common, as pairs joined others. As with the flying birds there was a strong tendency towards even-numbered groups, reflecting the preponderance of mated pairs.

The observed grouping can be tested more formally through nearest-neighbour analysis, by estimating the distance between each male and its nearest neighbouring male in the dunes. Most Ythan males had another within 10 m (figure 5.4) and most of these were within 5 m. The nearest-neighbour distances expected by chance, given the known overall density in the whole nesting area, were generated by a computer programme and this showed that the shelduck pairs were crowded together much more than would be expected (figure 5.4). I also observed pairs landing in the area and found that about 60 per cent landed beside other shelducks (with no consistent difference whether

Table 5.2. *Percentage of landings which were beside other shelducks. Sample size in brackets. From Patterson & Makepeace (1979)*

	Meadows	Dunes
1975	58.2 (67)	40.0 (5)
1976	59.3 (59)	67.6 (34)
Mean	58.7 (126)	64.1 (39)

None of the differences is statistically significant.

the birds were landing on the dunes themselves or in the meadows or dune slacks between them; table 5.2). It is difficult to calculate the proportion of birds which would land beside others if they came down randomly but some estimate can be made from the number of occupied and unoccupied areas available. In both 1971 and 1972 there was a maximum of 13 separate groups of shelducks (and usually fewer) on any one count of the Ythan dune study area, in which there were at least 50 places where groups had been seen at one time or another. Thus, on any one occasion, at least three-quarters of these places were unoccupied. Despite this, most shelducks landed in the minority of occupied sites and so were apparently selecting places already containing others. Hori (1964a) also recorded that some shelducks 'appeared to be drawn down, changing flight direction when they observed a flock on the marsh. Other birds appeared to search the grazing marshes for pairs and groups with which to associate', although these observations were made early in the season on preliminary gatherings away from the main nesting areas. Young (1970b) noted that a group forms 'when a single pair of birds arrives on the nesting grounds . . . As other pairs arrive on the scene, these are attracted by the birds already on the ground and they alight in the same place.'

Although such gathering into groups is almost certainly a social response to the presence of the other birds, it is possible that each pair responds independently to some feature of the habitat, such as a high density of nest holes. The formation of groups in slightly different places on different days, depending on where the first pair landed, makes this less likely, but the possibility can be excluded formally with a decoy experiment.

Table 5.3. *Responses of shelducks to an experimental group of decoys.*
From Patterson & Makepeace (1979)

	Area currently with	
	decoys[a]	no decoys[a]
Number of birds hovering	77	2
		$\chi^2 = 43.6, P < 0.001$
Number of birds landing	49	0
		$\chi^2 = 30.1, P < 0.001$

[a] The decoy and control areas were used alternately. The χ^2 tests were calculated against an expectation of equal numbers in the two areas if there was no effect of the decoys.

Decoys can be made by removing the viscera and major muscles from any shelducks found freshly dead, injecting the bodies with concentrated formalin and wiring them into the required posture until dry. I used a group of three pairs with two males in an Alert posture, two females with necks extended into nest holes and the remaining two in resting postures. The decoys were grouped around nest holes in the centre of a 20 m length of dune ridge overlooked by an observation hide in April 1977. An equivalent area, 20 m further along the same ridge, was used as a control and the decoys were exchanged between the two areas on alternate tests. The decoys were positioned soon after dawn, watched continuously for up to three hours and removed at the end of the session. Almost all of the shelducks hovering overhead, and all of those landing in the two areas, did so in the one currently occupied by decoys (table 5.3). Since the experimental (decoy) area and control area were exchanged on alternate days and were almost identical sections of dune, the birds must have been reacting to the presence of the experimental group, set up arbitrarily on an area not normally used by shelducks (confirmed by the low number occurring in the control area). This experiment thus confirms that shelduck groups in the nesting area are produced by a social attraction of flying birds to others already on the ground.

5.3 Sub-groups: 'parliaments' and 'communes'

Gillham & Homes (1950), following Coombes (1949), referred to groups of shelducks in the nesting area as 'the familiar parliaments', perhaps noting their similarity to displaying groups of rooks *Corvus frugilegus*, traditionally given this name. Young (1970*b*), in a paper

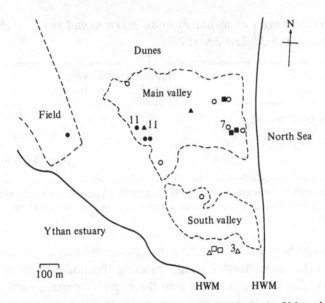

Figure 5.5 The dispersion of six marked pairs in the Ythan dunes,
March–June 1972. Each type of symbol refers to one pair and the
figures show the number of sightings at that point. Modified from
Patterson & Makepeace (1979).

entitled *Shelduck Parliaments*, used the term for consistent gatherings of
the same ringed individuals at the same site day after day, continuing for
some weeks. The term, with its legislative overtones, was consistent
with Young's ideas of socially induced exclusion from breeding which I
will discuss in section 6.8. Hori (1964*a*, 1969) also recognised the
consistent association of the same individuals at the same places in the
nesting areas throughout the season and termed these groups 'communes', 'colonies of breeding adults' and 'persistent groups of adults in
the nesting areas'. He found that such groups formed in early April and
continued until all the pairs in the group had either failed or had hatched
their broods. Although 'commune' implies fewer group activities than
'parliament' it is probably best to avoid such terms, with their possible
impression of the same functions as human institutions of the same
name, and I prefer, and will use, the more neutral term 'sub-group' (of
the local population).

Shelducks in the Ythan nesting area concentrated in particular parts
of the nesting habitat, especially in certain meadows between the dunes,
preferring those containing a pool. Many sites were around the peri-
meter of a large valley between major dune ridges, but one was on the

Table 5.4. *The positions of ranges (most used area), within the Ythan nesting area, in relation to the order in which birds were first seen. From Patterson & Makepeace (1979)*

| Order of arrival | Number of pairs with ranges in each part of area | | |
	West	South	East
1971			
1st 6	**3**	2	1
2nd 6	2	**3**	1
3rd 6	**3**	1	2
last 7	0	0	**7**
1972			
1st 6	**4**	2	0
2nd 6	**4**	2	0
3rd 6	0	**3**	**3**
last 7	1	0	**6**

For each set of birds, the most commonly used area is in bold. In 1972, first 12 birds versus the rest, west versus south plus east, Fisher exact test, $p = 0.03$.

sea beach and one was in a field between the dunes and the estuary (figure 5.5). Pairs and small groups also occurred throughout the dunes and minor valleys. Sightings of those marked birds seen more than once showed that most had restricted ranges within the dunes, each pair mainly being seen at one site, with other records usually at nearby ones. Each site was thus attended by a group of pairs with similar ranges, although most pairs also visited adjacent sites, as Hori (1969) also found, particularly when the birds were disturbed. Visits to areas more than a few hundred metres away were, however, rare among the Ythan birds. Marked pairs seen in the northern part of the sand dune system were seen only exceptionally in the main study area 3 km to the south.

Sites in different parts of the Ythan nesting area were settled in order of their closeness to the estuarine feeding grounds. Marked pairs were divided into groups in order of the date on which they were first seen, and each pair was assigned to the east, west or south of the study area, according to which site they used most (figure 5.5). In both 1971 and 1972 the earliest pairs were those using the west part of the area nearest to the estuary and the last pairs were generally those using sites in the east (table 5.4).

Individual pairs mostly used the same area within the nesting habitat in successive years. Hori (1969) showed that many ringed females used the same nesting locality in different years. All but one of 20 ringed

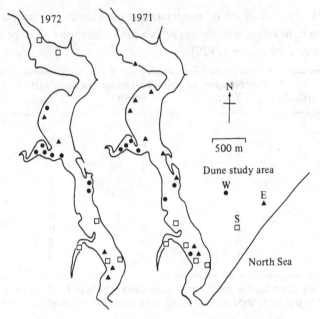

Figure 5.6 Territory sites of marked pairs with ranges mainly in the west (W), east (E) or south (S) of the Ythan dune study area. From Patterson & Makepeace (1979).

Ythan pairs returned to the same part of the dunes in successive years in 1970–2 (Patterson & Makepeace, 1979). This is a significantly higher degree of site tenacity than would be expected if the birds changed randomly around the three sections available ($\chi^2 = 8.88$, $P < 0.01$). Two females which changed their mates between years also returned to their previous areas, but one male with a new mate changed his area. These few cases do not allow any conclusion but are consistent with the female determining the nesting area to be used, as was suggested for the territory in section 4.4.

There is a tendency for shelduck pairs using the same area of the nesting habitat to have feeding territories which are close together. Hori (1964a) found that three ringed pairs in one sub-group had adjacent territories and that two pairs in another group were also territorial neighbours. In the Ythan population, the pairs using sites in the south of the study area mostly had territories on the adjacent southern part of the estuary (except for two pairs in 1972, which had territories far upstream; figure 5.6). The pairs using mainly the west dunes had territories in the mid-part of the estuary, including a group of four or five adjacent territory holders. The pairs using the east dunes had territories scattered more widely than the other categories (figure 5.6).

There is thus clear evidence that shelduck breeding populations contain consistent sub-groups. The same individuals associate in the same part of the nesting grounds morning after morning, they are often territorial neighbours and usually return to the same area in successive years. There is likely to be a high degree of individual recognition among such small sub-groups, an excellent basis for the formation of dominance relationships, which I will discuss in the next section.

5.4 Behaviour in the nesting area
General behaviour

Having flown to the nesting grounds at dawn, the female shelducks begin to visit holes and other potential nest sites. Some searching, or prospecting, for these is done in flight as the pair circle slowly and repeatedly around the area, usually with the female leading. Hori (1964*a*) described an attractive spectacle of pairs flying around and between trees containing nest holes. The birds land frequently and the female visits holes, peering into them and sometimes entering one partially or completely before moving on to another. Some of the holes entered can be quite small and I have seen females struggling to push themselves in. They all, however, reappeared head first, having found space to turn round inside. The time spent in the hole can range from a few seconds to several minutes. Hori (1964*a*) heard soft croaking quacks given by females while prospecting but they are often quite silent. It is difficult to determine when a hole has been selected for nesting and it is likely that a preference may gradually develop over the prolonged prospecting period of several weeks. Hori suggested that at holes subsequently used as nests, the females 'spent rather longer in the holes and then flew out in a more purposive manner, often flying a hundred yards or more and causing their mates to fly after them'. However, he does not present evidence to show that such females were the same individuals which subsequently nested in these holes. The prolonged stay could reflect the suitability of the hole as a potential nest site, leading to prolonged investigation. I have also seen the same abrupt flying away from holes which were not subsequently used for nesting.

The male usually plays little part in the actual investigation of holes. At most he accompanies the female around the area, remaining alert and sometimes standing on a high point while she walks from hole to hole. During prolonged bouts of prospecting, especially by several pairs together, the males often retire to an open area and preen or sleep till the females rejoin them. I have only rarely seen a male looking into holes at the Ythan but Hori (1964*a*) described much more active involvement of the male in 'many instances . . . where females appeared inexperienced or excessively apprehensive. In these the male attempted to induce his partner to inspect trees more closely by leading her to them with a variety of displays . . .' These included Head-throwing, apparently directed towards the nest sites, 'leading' of the female by walking in front of her with the neck extended parallel to the ground and the bill pointing vertically downward and 'enticing' by a posture similar to Head-down or female Inciting, with rapid scooping movements of the bill. Females responded to these displays by actively investigating holes in the vicinity. No similar male activity has been seen in hundreds of bouts of prospecting by Ythan pairs. Head-throwing by males was common but chiefly occurred as a mutual display between males in groups, or was shown by a bird on the ground as others flew nearby, so that it could reasonably be interpreted as a threat (as it normally is in the winter flock and on territory). I have seen a few instances where a male has extended his head and neck in Hori's 'enticing' posture, but in each case the male's open beak was pointed at the female who either ignored the male or took a few steps away. The posture appeared to be a mild threat but its significance is obscure.

The proportion of time spent in different activities by prospecting shelducks can be measured more exactly by watching them from an observation hide and recording the behaviour of each bird at intervals. For the Ythan birds, a 5 min interval was used for those in the meadows between dunes and a 1 min interval for those actively prospecting on the dunes themselves, since their behaviour changed more rapidly. The birds' behaviour was divided into seven distinct categories: sleep, with the head resting on the back and the eyes shut; sit/stand, with the neck bent in a relaxed posture; preen; walk; alert, with the neck extended vertically; prospect, looking into and entering burrows; and display, any agonistic or other posture.

In such time-sampling data, the proportion of time spent by the average individual in each behaviour can be estimated by the proportion of birds showing each behaviour during the spot observations at 5 or 1 min intervals. There are, however, statistical problems in testing the significance of any differences, say between the sexes, because the accumulated data contain runs of non-independent observations. Successive checks of the same bird at 1 min intervals, for example, will not

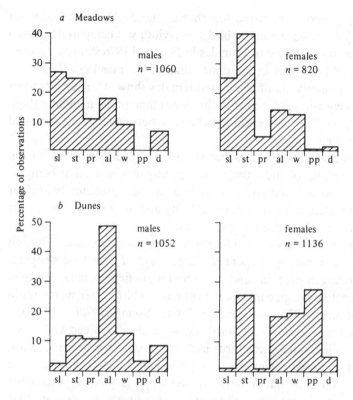

Figure 5.7 The percentage of observations made up by each type of behaviour in the Ythan nesting area. The types of behaviour (sl, sleep; st, stand or sit; pr, preen; al, alert; w, walk; pp, prospect; d, display) are described in the text. From Patterson & Makepeace (1979).

be independent of each other. For statistical testing, it is necessary to consider each marked individual as the basic unit of data; comparisons can then be made by counting the number of individuals which show mainly one kind of behaviour and the number which show mainly another.

Prospecting shelducks divide their time in the nesting area between active bouts of investigation of holes and longer 'rest' periods between, spent in open areas usually without any potential nest sites. The Ythan birds rested in meadows of short grass between the dunes, where both males and females spent most of their time asleep, sitting or standing (figure 5.7a). When on the dunes themselves, however, their behaviour differed considerably; the males spent most of their time alert while the females also stood, walked and prospected (figure 5.7b). (Walking and prospecting generally occurred together as the ducks walked between nest holes.) When each marked bird's behaviour was compared between meadow and dune, 14 out of 15 males in 1975 and 1976 showed a higher proportion of time alert in the dunes than in the meadows (Sign test, $P < 0.01$). Similarly, all of 18 marked females showed a higher proportion of walking and prospecting in the dunes than in the meadows (Sign test, $P < 0.01$). The differences in behaviour between habitats persisted throughout the season.

These observations confirm that the female does almost all of the active prospecting of sites, the male's principal contribution being to remain close to her and alert, possibly as an anti-predator behaviour since the birds enter areas of poor visibility and thick cover where they could easily be approached by predators.

While in the nesting area shelducks are surprisingly inactive. While on meadows, the Ythan birds spent 60–70 per cent of their time sleeping, sitting, standing or preening and they spent more time on meadows than they did in active prospecting (72 per cent of 369 birds, seen on counts in March and April 1972, were on meadows). Young (1970b) also commented that on meadows 'the birds do little else but sleep and preen . . .'. It is unlikely that these birds had already selected their nest sites, since most of them eventually prospected again later in the same morning or on subsequent days. However, it is possible that pairs developing a preference for a site were staying nearby to prevent other pairs from using it, or were assessing the frequency of visits by predators to the area. Possibly the inactive pairs were waiting for others to begin prospecting, so as not to be the first to enter potentially dangerous areas of thick cover. Others might have been deterred from prospecting by being joined by others, as I will discuss in section 5.5.

Aggression and dominance

A considerable amount of agonistic behaviour is seen among prospecting shelduck whenever two or more pairs come together. Hori (1964a) referred to 'the volume of display and bickering which occurs' among groups, and Young (1970b) described 'a good deal of calling and gesturing among the drakes . . .' The Ythan males spent over 5 per cent of their time displaying, more than the females (figure 5.7) and male–male encounters occurred in 29.2 per cent of 120 cases when two or more pairs were seen together. (This estimate is a minimum, since most groups spent some of their time out of sight of the observer and additional encounters could have occurred then.) Most interactions

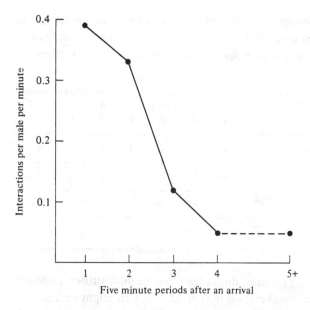

Figure 5.8 Frequency of aggressive interaction between males in groups in the Ythan nesting area, in relation to time since the arrival of a new pair in the group.

were supplanting attacks in the Head-down posture. The mean rate of interaction was similar in different years (the number per bird per minute was 0.040 ± 0.009 in 1970, 0.038 ± 0.013 in 1971 and 0.043 ± 0.009 in 1972) and did not change consistently with group size, habitat or date.

Interactions were much more frequent immediately after a new pair arrived to join a group (figure 5.8). Young (1970*b*) also noticed that display was more frequent as new arrivals came in. When a pair landed beside other birds, attacks by the arriving male towards those already there were no commoner than the reverse. There was no indication that males were attempting to defend particular areas, although in several cases a male appeared to attack others only when they approached his prospecting female. Occasionally a male attacked a female of another pair but almost all the interactions were between males. The significance of the male aggression is not clear, but it may, as in the winter flock, be concerned mainly with maintenance of the pair bond. Although virtually all the males in the nesting area are already paired, the mates of some may have already started to incubate, and such males could enhance their fitness by attempting to copulate with another

Table 5.5. *The dominance relationships between males in one sub-group in the Ythan nesting area, 1974. The data are the number of encounters won by the bird in the row over the bird in the column. From Patterson & Makepeace (1979)*

		Loser (bird number)						
		1	2	3	4	5	6	7
Winner (bird number)	1	—	5	8	2	—	4	9
	2	—	—	9	1	—	—	—
	3	—	—	—	8	7	2	2
	4	—	—	4	—	—	—	—
	5	—	—	—	—	—	3	—
	6	—	—	—	—	—	—	1
	7	—	—	—	—	—	—	—

female. The male's aggression might also ensure undisturbed prospecting by his female, as I will discuss in section 5.6. This might explain why the rate of interaction was highest when a new pair arrives, as each male attempted to drive the other away from his female.

By recording the identities of the marked males which won or lost each encounter, it became obvious that there were consistent dominance relationships between the Ythan males in the nesting area, as well as in the winter flock as I showed earlier in section 3.6. Seven males which were regular visitors to one dune meadow in 1974 could be placed in a linear dominance hierarchy (table 5.5) and very similar results were obtained for two other groups in 1976 and 1977. The relative dominance of males in the nesting area was the same as that found in the winter flock, for pairs of males seen interacting in both situations.

5.5 Interference between prospecting pairs
Reduction in prospecting

Since prospecting shelducks are attracted to others to form groups, in which aggressive interaction is common, it is possible that the birds might interfere with each other's activities. Certainly, prospecting activity by one female quickly attracts others in the same area. Groups in the Ythan dune meadows spent long periods without activity, but if one female walked or flew to a dune to visit nest holes, another female in the group frequently followed her within 1 min (79.8 per cent of 94 cases in 1975 and 1976; Patterson & Makepeace, 1979). The two females usually moved within a few metres of each other and frequently looked in turn into the same holes, also noted by Young (1970b). Some females interacted aggressively (in 21.7 per cent of 120 cases of at least two females together, 1975 and 1976; Patterson & Makepeace, 1979). They usually showed mutual Head-throwing (rarely seen in females away from nest holes) and occasionally there was a short lunge or peck between birds very close together.

Such following and interaction did not reduce the duration of a bout of nest-prospecting by Ythan females. The length of a bout was measured from the time a bird flew or walked on to a dune until she flew up or walked off it again. There was no significant difference in mean bout length between single females (5.2 ± 0.7 min, $n = 24$, 1975; 7.8 ± 1.4 min, $n = 31$, 1976) and females accompanied or later joined by one or more others (5.1 ± 0.8 min, $n = 33$, 1975; 12.1 ± 2.2 min, $n = 71$, 1976; Patterson & Makepeace, 1979). However, in 17.0 per cent of 112 cases where one female joined another already prospecting, the original female left the dune immediately (less than 10 s) after the arrival of the other. There was no consistent difference in the lengths of prospecting bouts when an interaction did or did not occur between females or pairs.

Although the length of a prospecting bout was thus not affected by the presence or aggression of others, the proportion of that bout spent actually investigating holes did change. Females accompanied by others spent a smaller proportion of their time prospecting compared with single females (figure 5.9). Most of this difference was due to females spending a lower proportion of time actually inside burrows when other females were present. Accompanied females spent more time standing and displaying than did single ones. These changes in behaviour when accompanied were shown separately by all of a sample of ten marked females in 1975 and 1976 (Sign test, $P < 0.05$; Patterson & Makepeace, 1979). Hori (1964a) also noted that 'the amount of real prospecting which such parties do is small compared with the volume of display and bickering which occurs'. Young (1970b) too found that 'As more and more birds join the group, the serious nest hunting gradually tails off and eventually ceases altogether, whereupon the birds usually adjourn to some plot of open ground such as a ploughed field or a grassy meadow between the dunes . . .'

Females thus seem to interfere with the prospecting of others by following them to holes and by reducing the amount of investigation of holes in each bout of prospecting. This may prolong the time which a

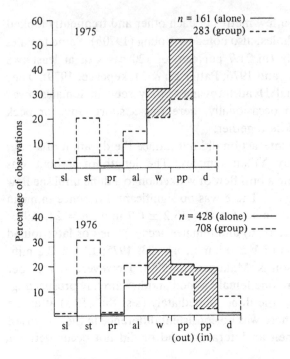

Figure 5.9 The percentage of observations made up of each type of behaviour in female shelducks in the Ythan dunes as a single pair (solid line) or in a group (dotted line). The types of behaviour are as in figure 5.7 except that pp (out) shows prospecting outside a burrow and pp (in) shows prospecting actually inside a burrow. The shaded portions indicate where single females show a higher percentage of a behaviour than females in a group. From Patterson & Makepeace (1979).

pair must spend in the nesting area and so may increase the risk of predation to which these conspicuous ducks are exposed when entering uneven terrain with thick cover.

Interference and density

It seems likely that interference between pairs will increase with density. It was possible to check this in the Ythan nesting area by using the number of different marked pairs seen from the observation hide as an index of the total number of shelducks present. The number of prospecting bouts where at least two females were present on a dune at the same time was used as a measure of the potential amount of interference on the same day. The number of such incidents of group prospecting increased significantly with increasing number of marked

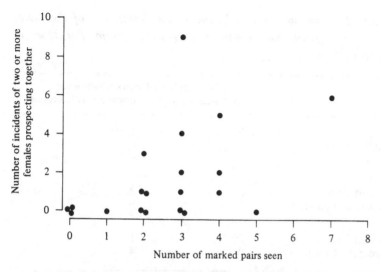

Figure 5.10 The frequency of incidents of two or more Ythan females prospecting together, in relation to the number of marked pairs seen that day ($y = 0.089 + 0.694x$, $r = 0.481$, $p < 0.05$). From Patterson & Makepeace (1979).

pairs seen in the area (figure 5.10). The increase was roughly proportional to the increase in the number of pairs seen, suggesting that interference may indeed increase with increased density of shelducks present in a nesting area.

Interference and nesting success

Does interference between pairs have any effect on their eventual success in hatching a brood? This question is difficult to answer directly because the burrow nests are often difficult to find and reach, and any investigation of them may cause the duck to desert her clutch (Young, 1964a). However, all territorial pairs are likely to start a clutch (Williams, 1973; Patterson *et al.*, 1974) and their hatching success can be measured by daily searches for newly hatched broods. Marked pairs prospecting the Ythan nesting area were divided into (a) those more often seen 'alone', i.e. as an isolated pair or initially isolated even if later joined by others, and (b) those more often seen 'grouped', i.e. having joined a pair already prospecting or having arrived on the dune in company with others. There was no tendency for pairs which were usually alone while prospecting to be more successful than pairs which were in groups, although the sample was really too small for adequate testing.

Table 5.6. *Nesting success in relation to the dominance of the male, within sub-groups in the Ythan nesting area. From Patterson & Makepeace (1979)*

| | Number of pairs which were | |
	seen with broods	not seen with broods
1974		
dominant 3 pairs	3	0
subordinate 4 pairs	0	4
1976		
dominant 7 pairs	5	2
subordinate 7 pairs	1	6
1977		
dominant 4 pairs	3[a]	1
subordinate 4 pairs	0	4

[a] The three most dominant pairs.
Fisher exact test; 1974, $p = 0.029$; 1976, $p = 0.051$; 1977, $p = 0.071$.

An alternative test, although an indirect one, involved dominance. Since the approach of one pair to others commonly involved aggressive interaction, the dominance relationships within the groups should allow the more dominant pairs to reduce interference with their prospecting, if it would benefit them to do so. They could, for example, keep others further away than could subordinates.

In three sub-groups in different years the more dominant males were more successful than the subordinates in hatching broods (table 5.6). In a smaller sample in 1975, only two of the marked pairs seen in a group were successful in hatching broods, but both had dominant males which won all their observed encounters with other males in the group. Dominant males were thus strikingly more successful than subordinates within the same sub-group. Although one possible reason for this difference may be that dominants should be less vulnerable to interference by others, there are of course other possible explanations. Dominants, for example, may be able to compete more successfully for 'better' nest sites, giving a higher chance of hatching success, or may have preferential access to other resources or may be generally fitter.

The question of whether interference with prospecting for nest sites has a causal effect on nesting success must thus remain open. However, if there is such an effect it raises interesting questions of how it might come about. It is possible, for example, that a reduction in prospecting efficiency might reduce success directly, or that the attendance of other females during the selection of a nest site might increase the chance of later interference by parasitic egg-laying. Jenkins *et al.* (1975) suggested that nest failures among a small sample of shelducks using nest-boxes at Aberlady were associated with the presence of new eggs laid by second females in the partly incubated clutches. Weller (1959) has shown that in the redhead *Aythya americana* nesting success was reduced when several females laid in the same nest. However, Hori (1964a) showed that in his area, eggs in multiple clutches of shelducks hatched as successfully as those in single ones. The question of the possible effects of interference would repay further study, since, as I will show below, there are important density-dependent effects on shelduck hatching success. There is also the intriguing question of why shelducks should form groups in the nesting area when this results in so much aggression and interference with each other.

5.6 Costs and benefits of grouping while prospecting

In many ways it is surprising that shelducks form groups in the nesting area. At the same time of season they are territorial on the feeding area, and their behaviour changes from aggressive territoriality to gregariousness within a few minutes as they fly from one habitat to the other. Their nests, being in holes and thus highly cryptic, should suffer least from predation if they were considerably dispersed (Tinbergen, Impekoven & Frank, 1967; Taylor, 1976) and the simplest way to ensure this would be for the pairs to avoid each other and space out while prospecting. Instead, however, they are attracted to each other and form groups, as I have shown. Such grouping results in a number of costs, already discussed, such as time and energy spent in aggression, loss of time for prospecting, the possibility of future interference or parasitic egg-laying and perhaps an increased risk of nest predation through nests being close together. Why then do the birds form groups? What are the benefits of grouping which might offset the apparent disadvantages or costs? At least two hypotheses can be proposed: (a) that grouping reduces the risk of predation on the adults, and (b) that some pairs may gain information on those nest sites which give the best chance of success in hatching a brood.

The anti-predator hypothesis

The various disadvantages of grouping may be offset by the anti-predator advantage of increased vigilance of a group and a conse-

Table 5.7. *Percentage of pairs and groups which flew up when approached by an observer. Sample size in brackets. From Patterson & Makepeace (1979)*

		Single pairs	Larger groups
a Meadows			
	1971	27.3 (44)	38.6 (57)
	1972	36.9 (65)	30.7 (75)
b Dunes			
	1971	33.3 (10)	69.2 (13)
	1972	37.5 (32)	54.2 (24)

None of the differences between pairs and groups is statistically significant.

quent increase in the likelihood of detecting a predator. Alternatively, there could be a decrease in the proportion of time each individual needs to be alert with a consequent gain of time for other activities (Murton, 1968). Pairs which join others also avoid the possible risk of being the first to land in an unknown area of thick cover. In a long-lived species like the shelduck a risk to the adult represents a greater potential loss of genetic fitness than does the same risk to one year's clutch of eggs.

This hypothesis predicts that birds in groups should detect a predator more quickly and individuals should spend less of their time alert than single pairs. All shelducks should show the same preference for landing beside others rather than in an unoccupied area and the female of a single pair arriving in an area should take longer to start prospecting than a female arriving to join a group.

Real predators, such as foxes and stoats *Mustela erminea*, are generally seen too rarely to be useful for testing the hypothesis. However, an observer following a concealed route and attempting to stalk the birds (to try to identify ring combinations) is probably an effective substitute. In the Ythan dunes, pairs and groups being stalked were divided into those which flew up in alarm and those which remained apparently undisturbed while being observed. The distance between the observer and the birds depended on the availability of vantage points and neither distance nor intervening cover could be controlled.

On the dune meadows, where the birds spent most of their time sleeping or sitting, there was no difference between pairs and larger

Table 5.8. *Percentage of observations in which males were alert, in relation to habitat and group size. Sample size in brackets. From Patterson & Makepeace (1979)*

		Single pairs	Larger groups
a Meadows			
	1975	16.5 (127)	21.2 (661)
	1976	25.0 (288)	15.3 (772)
b Dunes			
	1975	56.4 (117)	34.7 (167)
	1976	59.1 (492)	40.9 (560)

groups in the proportion which flew up (table 5.7a). On the dunes themselves, however, larger groups were about twice as likely to fly up than were pairs (table 5.7b). The difference was not statistically significant in either year or in all the data taken together but was fairly consistent in the two years. Single pairs were equally likely to fly up in both habitats.

Observations of behaviour at 5 min intervals, accumulated over all males, showed that on meadows there was no consistent difference in the proportion of time spent alert by males in single pairs and in larger groups (table 5.8a). On the dunes, the males in groups spent somewhat less of their time alert than did the males of single pairs (table 5.8b). This difference could not be tested statistically, being based on runs of non-independent data, but is confirmed by data on individual marked males. Each of eight males (in 1975 and 1976) was more often alert when alone with his mate on a dune than when he was with a group (Sign test, $P < 0.05$). Individual males on meadows showed no consistent differences in alertness with group size.

Females on meadows also showed no difference between groups and pairs in the proportion of time spent alert. They differed from males on dunes, however, by spending rather more time alert when in groups than when alone with their mate (figure 5.9), the reverse of the prediction of the hypothesis. Data from individual marked females showed no consistent differences in alertness with group size in either habitat. Females thus did not reduce the proportion of their time spent in alert behaviour when they were in groups and, as I showed in the previous section, they spent significantly less of their time prospecting when they were with other pairs. The females of single pairs arriving in an unoccupied area did not take significantly longer to begin prospecting (mean 13.9 ± 6.8 min, $n = 12$) than did females arriving in the company of others (mean 12.9 ± 3.7 min, $n = 10$; Patterson & Makepeace, 1979).

These results offer only limited support for the anti-predator hypothesis. Although groups on the dunes were, as predicted, somewhat more likely to fly up when approached by an observer (and so presumably by real predators) the difference was not significant. The males in dune groups showed the predicted lowering of the proportion of time they spent in alert behaviour but this difference did not extend to females, who of course do most of the prospecting. Since females in groups in fact spent less, not more, of their time prospecting, the benefit of the increased efficiency of predator detection probably did not result in any increase in the efficiency of nest-finding. The possible benefits of alert behaviour are not of course confined to defence against predators. The birds may also be watching for other shelducks (Jenkins *et al.*, 1975), which raises the question of the relationship between the prospecting pairs.

The nest-site hypothesis

A second possibility is that the various disadvantages of grouping, already discussed, are offset, at least for some individuals, by an increase in nesting success resulting from learning the location of good nest sites, e.g. those which have had successful clutches in the past. The net benefit of grouping will be greatest for females which do not already know a good site, i.e. females breeding for the first time and those which were unsuccessful in the previous year. Pairs which already know a good location could reduce the disadvantages of being joined by others by stopping prospecting when others appear or by driving them away. Dominant pairs should be most effective at this if it is in their interests to do so, i.e. if they have been successful in the previous year, which is likely (table 5.6).

This hypothesis, unlike the anti-predator one, predicts a difference in their tendency to join groups between females which hatched broods in the previous year and those which did not. The previously successful females should be seen less often in groups, should approach others less and should more often stop prospecting when others approach. Dominant pairs should more often be alone and should join others less than subordinates.

Table 5.9. *Proportion of marked females which more often landed in unoccupied parts of the Ythan nesting area. From Patterson & Makepeace (1979)*

Status of female	1975	1976	Total	Percentage
a Hatched brood in previous year	4/4	1/3	5/7	71
b Territorial in current year but did not hatch brood in previous year	1/6	1/5	2/11	18
c Non-territorial in current year	0/5	0/4	0/9	0

1975: *a* versus *b*, $p = 0.023$; *a* versus *c*, $p = 0.008$.
Both years together: *a* versus *b*, $p = 0.039$; *a* versus *c*, $p = 0.005$; Fisher exact test. None of the other differences is statistically significant.

Table 5.10. *Proportion of marked females which more often prospected alone in the Ythan nesting area. From Patterson & Makepeace (1979)*

Status of female	1975	1976	Total	Percentage
a Hatched brood in previous year	3/4	2/3	5/7	71
b Territorial in current year but did not hatch brood in previous year	2/4	3/7	5/11	46
c Non-territorial in current year	1/4	0/2	1/6	17

None of the differences between categories of female is statistically significant.

The marked females seen landing in the Ythan nesting area were divided into those more often seen landing in an unoccupied area (solitary pairs) and those more often seen landing beside others (group pairs). More solitary pairs were found among previously successful ones than among territorial pairs which had previously been unsuccessful (table 5.9). Females which were non-territorial in the previous year, and which had not bred before, all tended to land beside others.

Marked females which were already prospecting when first seen were also divided into solitary (those most often seen either alone throughout or initially alone and only later joined by others) and grouped (those already with other pairs and those which arrived in a group). Again, more grouped pairs were found among the previously unsuccessful or the previously non-territorial birds (table 5.10), although none of the differences was significant.

In the decoy experiment, described earlier, previously unsuccessful

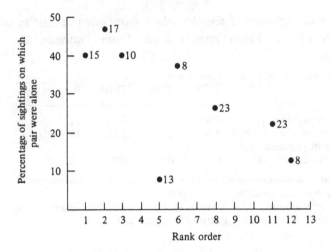

Figure 5.11 The percentage of sightings in which an Ythan pair were alone, in relation to the rank of the male. The figures are the numbers of sightings. Only pairs seen more than five times have been included. (Spearman rank correlation coefficient $rs = 0.719, p < 0.05$.) From Patterson & Makepeace (1979).

females landed beside the decoy group on more of their visits to the area and stayed with the group for longer than previously successful females. Although, again, the differences were not statistically significant, the trend was in the same direction as the previous observations. There was, however, no consistent difference between the categories of female in the time taken to stop prospecting when they were joined by others.

Pairs with males of known dominance in 1976 were similarly divided into those more often seen alone and those more often seen with others while prospecting. Four of five dominant pairs were solitary as compared with only two of five subordinate pairs. The difference is not significant but an even smaller sample from 1975 was consistent in direction (Patterson & Makepeace, 1979). The group size was recorded on the first time a pair was seen each day, irrespective of the habitat in which they were seen. The percentage of sightings in which the pair was alone was significantly higher among more dominant pairs (figure 5.11). This tendency for pairs with dominant males to be solitary was, however, not confirmed by the decoy experiment. The females of dominant males landed beside the decoys on more of their visits than did the mates of subordinate males, although again the difference was not significant (Patterson & Makepeace, 1979).

There is thus some support for the nest-site hypothesis, in that the predicted differences between categories of female were found. Previously unsuccessful females, which would have the greatest net gain from approaching others, did so. Dominant pairs also showed the predicted tendency to be alone more commonly than subordinates. This may have been due to driving others away rather than an avoidance of groups, since the females of dominant males showed no tendency to avoid the decoys. (This should, however, be regarded as a tentative conclusion, since the difference was not statistically significant.)

There seems to be rather stronger support for grouping as a means to gain information about nest sites than as an anti-predator behaviour (which I had initially regarded as the more likely!). The anti-predator hypothesis cannot easily account for the differences in grouping tendency between females of different previous breeding history. On the other hand the nest-site hypothesis may be able to account for the greater alertness of the male of the single pair, which may be watching for other shelducks in order to join, avoid or drive them away, depending on his dominance and previous breeding status. Drent & Swierstra (1977) suggested that variations in alertness in barnacle geese *Branta leucopsis* were correlated with variations in food abundance, and that the birds were watching other flocks. The increased anti-predator vigilance of larger groups could be a beneficial additional consequence of a basically social alertness.

There is of course no reason why the two hypotheses cannot co-exist. Previously unsuccessful pairs could obtain a dual advantage from grouping with others, while for previously successful pairs the disadvantage of being joined by others some of the time could be reduced, to some extent, by lessening the risk from predators.

5.7 Nest sites

Shelducks' nests are difficult to find and observe. Occupied holes are usually scattered among large numbers of unoccupied ones and the nest may not be visible from the outside, making it unproductive to search for nests in most areas. At Sheppey, however, where the birds used a more limited range of sites, many of them used in successive years, Hori (1964a, 1969) was able to find a large proportion of the nests in his study area. Usually the most practical way to find nests is to watch the female returning, but this is time-consuming, especially as the bird will often not enter the nest if she detects the observer (Hori, 1964a). Sometimes shelducks can be induced to use prepared sites and nest-boxes (see below).

Even when the nest is found there are difficulties, since the site may be inaccessible unless major modifications are made to the hole. In addition, many female shelducks are intolerant of disturbance and will desert the clutch if the hole or nest are altered. Hori (1964a, 1969) found that while some birds tolerated major changes to the nest site (e.g. demolition of hay stacks) and being caught on the nest, others were more timid and repeatedly deserted their clutches. Young (1964a) also found that many clutches which he investigated were subsequently abandoned. There may well be differences between populations in the degree of tolerance of human disturbance, depending on the normal amount of contact with man. On Sheppey, where many of the shelducks nest in close contact with man, or even in man-made structures, the birds may develop a greater tolerance.

As a result of these difficulties in studying shelduck nests there is a shortage of direct evidence, and data on such basic variables as clutch size and hatching success are remarkably scarce for such a common and well-known species. Throughout the laying and incubation periods a number of indirect methods have had to be used to overcome the limitations on direct observation.

The most commonly reported nest sites of the shelduck, at least in the western part of its range, are disused rabbit *Oryctolagus cuniculus* burrows or similar holes in the ground. All the nests found at the Ythan were in rabbit holes, as were most of those found at Aberlady, and the same is probably true of most sand dune and shore nesting areas. Gillham & Homes (1950) reported nests in pipes made in mounds on the salt marsh and intended as refuges for hares *Lepus europaeus*. Some of the Aberlady birds nested inland in cavities in stacks of bales (Jenkins *et al.*, 1975) and such sites were used commonly on Sheppey (Hori, 1964a). There, stacks were used for 25.5 per cent of 110 nests, while 24.5 per cent were in rabbit holes, 26.4 per cent in tree holes, and 2.7 per cent in open sites. The remainder were in a variety of places, including derelict buildings. About half of the tree holes were in the bases of the trees but the remainder were above ground, sometimes in hollow limbs 3–5 m up. Nests in hay and straw stacks were made in narrow tunnels between the bales and could be up to 8 m above ground level.

Shelducks have been reported nesting on the surface in a number of places, particularly among thick bracken *Pteridium aquilinum* and bramble *Rubus fruticosus* on the Medway (Gillham & Homes, 1950), among 'scanty salt-marsh vegetation at the base of a derelict portion of a sea-wall' and among cord grass *Spartina maritima* (Gillham, 1950), under massive hogweed *Heracleum sphondylium* (Kennedy, Ruttledge & Scroope, 1954) and even 'under a single paper meal bag' (Hori, 1969). I have seen a large number of nests in thick herbaceous *Salicornia* in the Camargue. In some of these surface sites the cover is so thick that the birds are effectively nesting in tunnels similar to underground burrows, but the more open sites are probably exceptional, reflecting a shortage of suitable holes. For example, Hori's nest under a bag was found in a year when there was a shortage of haystacks. Most of such reports of surface nesting and haystack sites come from areas where much of the land is low-lying and wet, leading to a shortage of burrows. There may, however, be differences between local populations in their nesting traditions if, for instance, females prefer the type of site in which they themselves were hatched.

Jenkins *et al.* (1975) made nest boxes for shelducks, some adapted from existing rabbit burrows and some newly constructed. These were not used much in the first year or two after construction, and those made in previously occupied burrows were usually avoided. Eventually, however, about one third (of 47) were used, greatly aiding study of the nests. Such nest boxes should be most successful in areas where natural cavities are scarce or can be blocked.

Within the nest hole most shelduck nests are fairly close to the entrance; the majority of the Ythan nests were within 1 m and few were beyond 2 m. Hori (1964*a*), however, found nests in bales to be up to 5 m from the face of the stack. Many holes have a bend in the tunnel so that it is not possible to see the incubating duck from outside. No nest as such is built, although the female usually excavates a shallow scrape if the substrate is soft (Hori, 1964*a*).

Dewhurst (1930) considered that all shelduck nests had a 'bolt hole', a second exit which could be used if a predator entered the main hole. Hori (1964*a*) also considered such escape holes to be 'a general requirement', and found them in three nests he opened up in rabbit burrows. However, he found that many such escape holes did not open to the outside but were blind tunnels in which the female could hide when disturbed (Hori, 1963). Sometimes the 'escape hole' was merely a continuation of the same burrow beyond the point where the bird had made its nest. It is not clear whether shelduck females actually select burrows with escape holes or whether most burrows have these in any case. It may be that the birds avoid nesting in short, blind holes where they could easily be trapped by a predator. Nevertheless, I have seen some nests in such burrows.

The amount of re-use of shelduck nest sites in successive years varies between areas. In the Ythan nesting area, of 35 sites found in 1962–4, only three were used twice (by different birds) and two others showed signs of having been used before, with old down and egg shells below the current nest. At Aberlady, 78–85 per cent of nests in the dune area were in new holes each year, whereas at inland sites about half of the sites were used in two out of three years (Jenkins *et al.*, 1975). At Sheppey many of the same sites were used every year, often by the same birds (Hori, 1963, 1964*a*, 1969). The amount of re-use of sites probably results from the number of holes available in relation to the size of the population. Where rabbit holes are very abundant, as at the Ythan (Williams, 1973), it may be advantageous to use a new hole each year. By changing, the birds will avoid any build-up of parasites and may prevent predators from learning the location of habitually used sites. On areas like Sheppey sites are probably much scarcer, so that the birds will be obliged to use the same places year after year. This has allowed local farm workers to remove eggs from clutches (Hori, 1964*a*).

6

Laying and incubation

Having selected a nest site, the breeding bird must build a nest, lay a clutch of eggs and incubate them. The number of young hatched will depend on the size of clutch which can be produced and success in hatching them, in the face of many factors which might cause failure. Incubating birds must divide their time between the eggs and the need to spend some time feeding, the balance depending on the fat reserves which can be lost over the incubation period. In this chapter I will discuss these and other problems encountered by laying and incubating shelducks.

6.1 The timing of laying

The date on which the first egg of the clutch is laid can only rarely be determined by direct observation during the egg-laying period, due to the inaccessibility of the nests and the birds' intolerance of disturbance. Instead, a number of indirect methods must be used. In a few nests, observed before laying is complete, the laying date of the first egg can be back-dated since normally one egg is laid per day (Hori, 1964a). If the hatching date and clutch size are known, the laying date can be estimated using the incubation period of 29–31 days (Hori, 1964a; Young, 1964a), plus one day for each egg in the clutch. More commonly the laying date is back-dated from the first sighting of the brood of ducklings, adding a further day which the young spend in the nest after hatching (mean of four nests observed at hatching by Young, 1964a). Hori (1964a), using this technique, added three days to allow for the age of the ducklings, presumably because he did not always see broods on their first day after emergence. Any delays in detecting broods and any losses of eggs or ducklings will of course introduce errors in the estimate of the laying date. Furthermore all back-dating from hatching and broods is restricted to successful clutches, which may not be laid at the same time as those which fail (see section 6.6), so introducing a bias.

In an attempt to overcome this problem, Williams (1973) developed a new method, using the fact that the male remains alone on the territory while the female is laying and incubating (discussed in section 4.7). In over 10 000 sightings of Ythan territorial birds during the nesting period, marked females were virtually never seen apart from their mates, except when they were on the nest. The technique was assessed for each successful pair by comparing the estimated laying date (from the brood's emergence date) with the date on which the male was first seen alone on the territory. In 1971, visits were made four times each day to 14 successful pairs which had easily observable territories. In these the males were seen alone on average 2.1 ± 0.83 days before the estimated laying date, with a range from eight days before laying to three days afterwards. Since females are quite likely to visit the nest in the days immediately before egg-laying begins, Williams concluded that sighting of the male alone gives an accurate estimate of the date of laying, provided that the territories are observed frequently.

Clutches are started over an extended period, from 25 April to 19 June on Sheppey (Hori, 1964a), and from 20 April to 12 June on the Ythan (Patterson *et al.*, 1974). There generally is no sharply defined peak of laying, but most clutches were started in early May with a long tail extending into June (figure 6.1). The females laying late in the season were almost certainly not re-laying after failure since none of the marked Ythan females which lost their clutches showed any evidence of starting a new clutch (Williams, 1973). Hori (1964a) also found no evidence of re-nesting. There is no evidence that the more northern Ythan population lays later than the Sheppey one. (Apparently later peaks shown for Sheppey are probably due to Hori's (1964a) presenta-

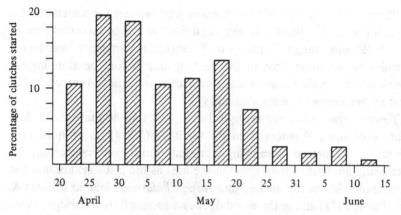

Figure 6.1 Distribution of dates when the first eggs of 122 Ythan clutches were laid in 1962–4 and 1971. Drawn from data in Patterson *et al.* (1974) and Williams (1973).

tion of the number of females laying at any one time rather than the more conventional number of clutches started on each date.)

Williams (1973) found that in pairs which occupied their territories earlier in the season the females generally laid earlier ($r = 0.56, n = 43$, $P < 0.001$). This suggests that a certain period on territory is necessary before egg-laying can begin, possibly to allow the female to build up her reserves.

6.2 Behaviour during laying

While laying, the female is typically absent from the territory for a single short period each day. While she is at the nest the male may return alone to the territory (Buxton, 1975) or may stay near the nest, perhaps joining other members of his nesting sub-group (Hori, 1964*a*). Buxton (1975) recorded two females which laid in the morning (0508–0847 h and 0858–1014 h) and one in the afternoon (1600–2020 h). Hori (1964*a*), however, considered that egg-laying normally occurred during the morning period when most pairs visit the nesting grounds and that laying later in the day happened when females were prevented from reaching the nest through human disturbance. One female laid her eggs at various times between 0645 and 1600 h, and another made at least four visits to her site before being able to lay at 1330 h.

Hori (1964*a*) found that egg-laying females usually spent 20–30 min in the nest, but one bird was found to lay and leave the nest in 6 min, apparently without being disturbed. The Ythan females, however, spent much longer periods away from the territory during egg-laying.

Williams (1973) found that the females were absent for a mean of 2 h 18 min (range 55 min–3 h 45 min) and Buxton (1975) recorded a mean of 2 h 20 min (range 30 min–4 h 49 min). This suggests that laying females spend some time in the nesting area before or after laying, which could enable them to make sure they were not observed at the nest by predators or other shelducks.

Towards the end of egg-laying Buxton (1975) found that females had more prolonged absences from the territory (5 h 41 min–8 h 10 min) which he interpreted as showing a gradual build-up of incubation. An increasing amount of time spent in the nest as the last eggs are laid has been shown in the red-necked phalarope *Phalaropus lobatus* (Hilden & Vicolante, 1972) and in the wood duck *Aix sponsa* (Breckenridge, 1956; Grice & Rogers, 1965). Some time will also be necessary for stripping of the female's belly down, which is packed around the eggs at the onset of incubation. Hori (1964a) found that down was added to the nest only at the last egg-laying visit. He never found any incomplete clutches with down (except for one which was less than one metre from an already incubating female). Partially completed clutches were, however, covered by other materials such as hay, straw, leaf mould or wood litter (in 51 per cent of 43 clutches observed).

6.3 Clutch size

Estimation of the number of eggs laid by individual females is complicated by the possibility of more than one bird laying in the same nest. Hori (1964a) recorded several instances where at least two females visited the same nest and laid eggs at different times of day. At one such nest one of the birds continued to lay after the other had begun incubating, until there were 25 eggs in the nest. The second female was seen trying to enter the site while the first one was sitting. Hori considered that the eggs of different females usually could be distinguished by differences in size, shape and colour. Jenkins *et al.* (1975) found that in some deserted nests, the eggs were at widely differing stages of development, suggesting that some had been laid after incubation had started. However, it is not always possible to distinguish the eggs of the separate contributing females by these criteria and such multiple nests must usually be identified by their size. Very large clutches, e.g. of 28 and 32 (Witherby *et al.*, 1939) and of 50 (Dementiev & Gladkov, 1952) are fairly obviously produced by several females since the maximum recorded from a single bird, even when its eggs were taken away as they were laid, was 18 eggs (with one second-hand report of 25; Hori, 1964*a*). In a sample of 60 nests at Sheppey, where the eggs were counted daily and the behaviour of the adults was observed, all clutches greater than 12 eggs were proved to involve more than one female (Hori, 1969). A further 69 nests in fairly solitary sites, with no evidence of visits by additional females, had no clutches exceeding 12 eggs. It thus seems likely that shelduck clutches with more than 12 eggs are multiple ones. (The reverse, however, is not necessarily true, since a clutch of fewer than 12 eggs could be made up of two small clutches.)

Multiple nesting probably results in two otherwise puzzling phenomena, deposition of both eggs and down away from the nest. Hori (1964*a*) found occasional eggs on the marshes and tidal flats, and Young (1964*a*) found two eggs on the nesting area and one on the Ythan estuary. Jenkins *et al.* (1975) found four fresh eggs in the entrance of a burrow in which a clutch had been incubated for two weeks. Hori (1964*a*) suggested that such eggs might be deposited by females scared from their own nests by human disturbance, but it is also possible that the second female laying in a multiple nest may be prevented from laying there once the first female begins to incubate continuously.

Down-stripping away from the nest has often been described. Gillham (1951*a*) found clumps of down on the marshes of North Kent from 27 May–26 June in 1949 and 1950, usually in places where groups of shelducks were seen. Young (1964*a*) reported down in the nesting area at places where groups assembled to sleep and preen. Hori (1969) also found down in the nesting areas and showed that it was produced by breeding females rather than the non-breeders suspected by Gillham (1951*a*). Hori (1969) found that most down was deposited two to three weeks after the peak laying date and considered that it was produced by females who had been prevented from incubating in multiple nests. He suggested that the amount of down found each year might be used as an index to the degree of multiple nesting each season.

The distribution of clutch sizes varied between the three areas where they were measured (figure 6.2). Most clutches were of eight (Ythan), eight to ten (Sheppey) or ten eggs (Aberlady), but Sheppey and Aberlady had many more clutches over 12 eggs than the Ythan population. Only Sheppey had clutches of over 20 eggs. The mean of single clutches (taken as those not exceeding 12 eggs) was somewhat higher at Aberlady, with 9.73 ± 0.43 ($n = 15$, Jenkins *et al.*, 1975) compared with 8.95 ± 0.17 ($n = 95$) at Sheppey (Hori, 1969) and 8.10 ± 0.23 ($n = 29$) at the Ythan (Young, 1964*a*) although none of the differences is statistically significant. The overall mean of single clutch-

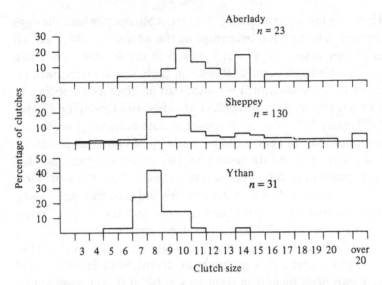

Figure 6.2 Distribution of clutch sizes in different areas. Drawn from data in Jenkins *et al.* (1975), Hori (1969) and Young (1964*a*).

es, taking all three studies together, is 8.85 ± 0.14 ($n = 140$). Later data from Aberlady (Pienkowski & Evans, in press *b*) gave a mean clutch of 8.94 ± 0.41. The mean clutch size in different years varied from 7.6 to 9.3 between 1962 and 1968 at Sheppey (Hori, 1969) and from 7.6–8.7 between 1962 and 1964 at the Ythan (Young, 1964*a*; Patterson *et al.*, 1974).

The proportion of multiple clutches (taken as those exceeding 12 eggs) varies between areas and between years (table 6.1). At the Ythan, Young (1964*a*) found only two in 31 clutches he examined. One of these, of 14 eggs, was in one nest but the other was a case of two nests less than a metre apart in the same tunnel, so that the eggs of the outer nest became scattered and mixed by the passage of the inner female. This nest should not be regarded as a true multiple clutch so that only one (3.1 per cent) of the 31 Ythan clutches was a combined one. The proportion of multiple clutches was significantly higher at Aberlady (34 per cent of 23 nests) and at Sheppey (27 per cent of 130) (χ^2 tests, $P < 0.05$), and these two areas were similar to each other (Hori, 1969; Jenkins *et al.*, 1975). The proportion of multiple clutches at Sheppey varied widely between years, from 11 per cent (of 19 clutches) in 1965 to 48 per cent (of 25 clutches) in 1963. The high proportion in the latter year may have been due to the destruction of many of the birds' usual

Table 6.1. *Percentage of multiple clutches (those containing more than 12 eggs) in different areas. From data in Hori (1969), Jenkins et al. (1975) and Young (1964a)*

		Total clutches	Percentage of multiple clutches
Sheppey	1962	13	23
	1963	25	48
	1964	22	14
	1965	19	11
	1966	25	28
	1967	13	38
	1968	13	23
	1962–8	130	27
Aberlady	1971–3	23	34
Ythan	1962–4	31	3

nest sites through consumption of hay stacks in the hard winter of 1962–3, forcing the females to share the remaining nest sites. Hori (1969) also considered the high level of multiple nesting to be an adaptation to recovery after a hard winter, since a larger proportion of eggs hatched in multiple clutches than in single ones (see below). It is difficult to support this argument, however, since he shows (Hori, 1964a, his fig. 2; Hori, 1969, his table IX) that the number of adults present in his study area in April and May was not lower in 1963 than in 1961 or 1962 and the proportion of multiple nests was also high in 1967 (38 per cent of 13 clutches), which was not preceded by an unusually hard winter.

6.4 Behaviour during incubation

Once incubation begins, the shelduck female remains on the nest overnight (Hori, 1964a; Williams, 1973; Buxton, 1975) and makes only a small number of visits to the territory to feed during the day. Hori found that at Sheppey, males came to the nest hole and called to their females with a 'clear strong whistle', and in the first two weeks of incubation females came off the nest quickly to join the male. Later Hori saw males fail to call off their mates in spite of repeated efforts. Some females were seen to leave their nests by themselves, which was invariably the case at the Ythan, where I never saw males coming to the nest to call. This difference in behaviour between populations may be related to the greater human presence and disturbance noted by Hori (1964*a*, 1969) at Sheppey, where it may be advantageous for the male to indicate to the female that the coast is clear for her to emerge.

On the Ythan, the female calls repeatedly on her way to the territory and the male may fly to meet her. When the two meet they perform a brief greeting ceremony, with Head-throwing by the male and Inciting by his mate. After feeding for some time, females almost always go to water to bathe and preen briefly before returning to the nest. Any water retained on the brood patch may moisten the eggs and keep up the humidity in the nest, traditionally an important factor in the artificial incubation of duck eggs.

The male appears always to accompany the female back to the nest, following her when she takes off (Hori, 1964*a* and my own observations). The pair may circle once or twice near the nest but often the female approaches directly. Sometimes she lands a few metres away from the hole and walks to it; on other occasions she flies directly to the entrance and Hori (1964*a*) has seen females flying directly into holes in the face of haystacks, closing their wings at the last moment. Female shelduck are, however, very sensitive to any slight sign of disturbance around the hole. I have watched from a concealed hide as females returned to their nests and found that after I had made a brief visit to the hole, taking great care to brush out any footprints and to rearrange any crushed grass, the ducks took very much longer than usual to enter the burrow. Most veered off when about ten metres from the hole and circled several times before landing at a distance and approaching cautiously on foot. Some returned to the territory for some time before repeating the approach. Williams (1973) made similar observations. Some males turn back immediately for the territory after the female enters the hole, others circle once or twice first and sometimes land in the general area, to sit or sleep for some time before returning to the shore.

The behaviour of the female on the nest has not been studied, presumably because of the nest's inaccessibility. I have made a few observations on a captive female which nested under a pile of logs against an old greenhouse. She spent much of her time in a resting posture, sometimes sleeping and occasionally preening. At intervals she stood up and poked with her beak among the eggs, which could be heard banging vigorously together. Hori (1964*a*) recorded 'widespread' damage to eggs which he attributed to stony ground and hard projections during turning and settling movements by the female. He found hairline cracks and small dents in some eggs and sometimes in whole clutches.

If the female is disturbed during incubation she will usually leave the nest and hide further along the nest tunnel or in a branch from it (Hori, 1963). Many females hiss loudly when disturbed, possibly to alarm a predator. Hori (1964a) saw a magpie *Pica pica* recoil violently from an occupied shelduck nest site possibly in response to the hiss. I have similarly seen a female shelduck, entering a burrow known to be occupied by another, rush out rapidly and fly away. Hori (1963) found one unusually fearless female which attacked his hand with wings and beak when he attempted to examine her eggs.

The down stripped from the female's brood patch is arranged around and below the eggs in a thick pad giving effective insulation from the ground. Before leaving to feed, the duck covers the eggs. My captive did this by drawing down from beside the clutch and inwards towards the centre to give complete coverage. This must substantially reduce the rate of cooling of the eggs in her absence and may help conceal the white eggs from any predator looking into the hole.

Feeding bouts during incubation

The time and duration of periods spent off the nest by the female shelduck can be measured by two methods, observation at the territory and regular checking or automatic recording at the nest hole. Observation, usually from dawn to dusk (Williams, 1973; Buxton, 1975) is laborious and may underestimate the number and duration of visits if the female feeds away from the territory. Regular checking of the nest is also time-consuming and may disturb the duck, so that several attempts have been made to devise an automatic recorder. Hori (1964b) used a domestic beam scale with the nest placed on the weighing pan. The weight of the bird on the nest activated a switch which controlled a record made on a continuous chart. There were problems with drag between the nest and its surroundings, and it would be difficult to install such equipment into most shelduck nest sites. Williams (1973) used a treadle switch concealed on the floor of the nest hole just inside the burrow entrance, and I have used a spring-steel wire projecting from the burrow wall (figure 6.3); both devices were connected to a continuous recorder. A disadvantage of this type of detector is that it records merely the passage of the bird (or rabbit!) without distinguishing whether it was entering or leaving. However, the timing of the records,

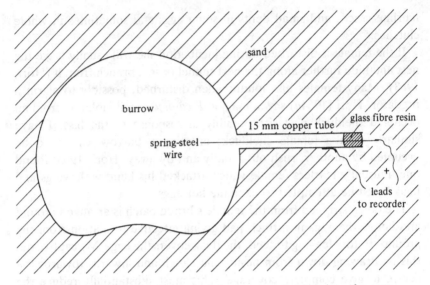

Figure 6.3 A switch to record the passage of a bird along the nest
tunnel.

the continuous overnight incubation and occasional checks at the
territory usually enable the recording to be interpreted.

Hori (1964a) found, in checks of nests throughout the 24 h, that the
female was always on the nest between 1800 h and 0400 h. During the
rest of the day he found the female off the nest on 32.6 per cent of 86
visits. Data from his automatic recorder showed that one female (unless
disturbed) left the nest only once on each of eight days, twice on another
eight days and left three times on one day giving a mean of 1.6 ± 0.2
times per day. She stayed off the nest overnight for 11 h on one occasion
(not included in the above). A female similarly recorded by Young
(1964a) came off the nest much more frequently, a mean of 4.5 ± 0.3
times per day, ranging from three to six times (n = 11 days). However,
other Ythan females, watched at the territory by Buxton (1975), came
off to feed on average 2.3 times per day, a figure much more similar to
Hori's. Williams (1973) found that a female he studied came off the nest
3.3 ± 0.3 time per day (range 2–5 times, n = 11 days).

The periods spent off the nest by the female are spread throughout
the day. Hori's (1964a) records show an equal number of departures
before and after noon, while Young (1964a) showed a slight preponderance of morning visits to the territory. Although shelducks normally
show a tidal rhythm of feeding, the female leaves the nest at any stage of
tide (figure 6.4).

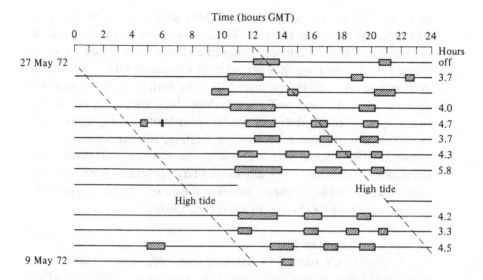

Figure 6.4 Pattern of incubation bouts for one female during days 14–27 of incubation. The shaded bars show periods off the nest and no line is shown when the recorder was not working. From Williams (1973).

The duration of stay at the territory varies considerably. Hori (1964*a*) found that the female he studied stayed off the nest for a total of up to 6.5 h during the day but he did not give an average time. Buxton (1975) found a group of females had a mean stay of 3.8 h per day on the territory while Williams' (1973) nest recorder showed an average period off the nest of 1.3 ± 0.1 h (range 0.1–3.2 h, *n* = 35), giving a total of 4.2 h per day (figure 6.4).

Individual female shelducks thus vary considerably in the number of breaks of incubation during the day and in their duration of stay while off the nest. It is possible that those which come off the nest less often stay for longer on each occasion, but too few individuals have been studied intensively for this to be tested. The time each female requires for feeding may vary depending on her energy output while incubating, her reserves and on the abundance and availability of food on the territory.

6.5 Hatching

Shelduck eggs hatch after about a month of incubation; Hori (1964*b*) found that one clutch took 31 days to hatch, nine others took 30 days and two clutches incubated by domestic hens took 29 days, giving a mean of 29.9 ± 0.14 days. Young (1964*a*) gave a mean of 28.8 days (range 26–31) for five nests at the Ythan. There can be a wide spread in the hatching dates of individual eggs in a clutch. Hori (1964*a*) found a completely dry duckling in a nest with 11 eggs, only four of which had even started to chip. He estimated that the earliest duckling was two days ahead of the next and possibly four days before the last of the clutch. Several other nests had similar spreads of hatching dates. Such variation may be caused by multiple nesting, so that some eggs are part-incubated when others are laid. Some eggs may be delayed by inadequate incubation, especially in very large clutches where eggs may spend time buried below others or below the down. Once the eggs begin to chip, they take 24–36 h to hatch (Hori, 1964*a*).

6.6 Nesting success

Shelduck's success in hatching their eggs can sometimes be measured in the conventional way, as the proportion of the eggs which hatch. Usually, however, the nests cannot be observed and nesting success must be measured as the proportion of territorial pairs which produce broods. Losses among the successful clutches can be estimated by comparing the mean brood size with the mean clutch size in those nests which were observed. Any losses detected by this comparison could have occurred either as eggs which failed to hatch or as ducklings which were lost before the brood was first seen.

Hatching success of individual eggs

A high proportion of shelduck eggs hatch successfully. Hori (1969) found that in successful single clutches on Sheppey, 87–93 per cent of the eggs hatched (table 6.2*a*). The number of eggs on which the percentages were based are not given, so that a true overall mean cannot be calculated, but it is likely to be close to 90 per cent. Pienkowski & Evans (in press *b*) found that 92.4 per cent of 105 eggs hatched at Aberlady.

Perhaps surprisingly, multiple clutches were not less successful than single ones (table 6.2*b*) with 85–95 per cent of the eggs hatching in successful nests (i.e. those which did not fail completely to produce any ducklings). The overall mean is again likely to be nearly 90 per cent of eggs hatching. This finding contrasts with Weller's (1959) suggestion that multiple nests are less successful than singles, though it does not take into account the possibility that more multiple clutches might fail completely.

Table 6.2. *Percentage of eggs hatching within single and multiple clutches. From Hori (1969). Number of clutches shown in brackets (number of eggs not given by Hori)*

		1963	1964	1965	1966	1967
a	Single clutches	90 (8)	93 (12)	87 (9)	92 (8)	88 (6)
b	Multiple clutches (over 12 eggs)	91 (8)	95 (2)	90 (2)	89 (6)	85 (3)

Table 6.3. *Hatching success of clutches. From data in Jenkins et al. (1975) and Young (1964a)*

	Number of clutches			
Fate of the clutch	Aberlady	Ythan	Total	Percentage
Hatched	10	19	29	48
Deserted (natural causes)	9	2	11	18
Deserted (human interference)	4	14	18	30
Probably predated	0	2	2	3
Total clutches	23	37	60	

Hatching success of clutches

Young (1964a) found that on the Ythan, excluding nests which were deserted through human activities, 83 per cent of clutches hatched some ducklings (table 6.3), while Jenkins *et al.* (1975) recorded 53 per cent success. These may very well be overestimates since the clutches lost by desertion after human disturbance might have been those most likely to fail in any case. If such desertions are included, the percentages above are reduced to 51 and 44 per cent respectively.

Boase (1951) reported that only about half of the territorial pairs on the Tay were subsequently seen with ducklings. From counts of breeding pairs and the approximate number of broods represented among the ducklings seen on Sheppey (Hori, 1964a), hatching success of clutches can be estimated at 26.8, 26.0, 40.0 and 78.1 per cent in 1960–3 inclusive. Jenkins *et al.* (1975) reported an even lower success at Aberlady with a mean of 12.6 ± 1.6 broods (range 4–18) being seen

Table 6.4. *Hatching success of clutches on the Ythan 1962–79. From Patterson et al. (in press b)*

Year	Number of territorial pairs	Broods (expressed as a percentage of the number of territorial pairs)
1962	71	38
1963	70	31
1964	72	25
1970	62	39
1971	76	37
1972	72	28
1973	61	42
1974	56	55
1975	60	35
1976	63	51
1977	82	51
1978	70	53
1979	65	42
1962–79	68	40.5

over 7 years (1967–73), when about 50 pairs were thought to have nested. This represents around 25 per cent (range 8–36 per cent) hatching success of clutches laid. On the Ythan an average of 40.5 per cent of the territorial pairs were subsequently seen with broods (table 6.4). Success varied between years, from 28 to 55 per cent. Thus, although hatching success is high within successful clutches in the shelduck, a large proportion, usually more than half, of the territorial pairs fail to hatch any of their eggs. Success varied between years and between areas, although some of the variation may be due to differences in technique between observers. For example, the likelihood of seeing a brood which disappears soon after hatching will depend on the frequency of observation.

Timing of failure

Shelduck nests can fail at any stage from laying to hatching, but most failures probably occur early. Jenkins *et al.* (1975) did not see more than 26 males alone on their territories at Aberlady in an area where 40–45 pairs were nesting, and suggested that nesting losses may have occurred before or soon after incubation started. Williams (1973) estimated the period for which nests survived by the length of time for

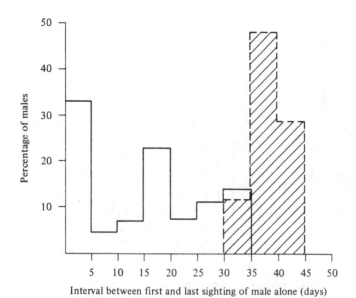

Figure 6.5 The period in 1971 for which males were seen alone on territory, in pairs which were successful (shaded histogram, dashed line) and those which failed (open histogram, solid line). Re-drawn from Williams (1973).

which males were seen alone. In pairs which hatched their broods, most males were alone for 35–40 days (figure 6.5), which corresponds well with the combined laying and incubation periods. In unsuccessful pairs, about a third appeared to fail within the first five days, i.e. during laying or very early incubation. Other pairs failed throughout the incubation period, with a peak 15–20 days after the male was first seen alone. The latest failures were 35 and 37 days after the first sighting of the male alone, by which time the eggs must have been very close to hatching or had actually hatched. All but one of the failures recorded by Williams (1973) occurred after 10 May, with a peak around 25–30 May.

Why do nests fail?

Williams (1973) found that successful shelduck females on the Ythan laid significantly earlier (as judged from when their males were first seen alone) than did unsuccessful ones (figure 6.6). Early females, those whose males were seen alone before 10 May, also had their failures later in the incubation period (males seen alone on average for 19.8 days) than did later females (males seen alone for 11.3 days). These

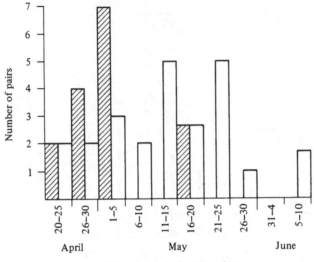

Figure 6.6 Laying dates of successful pairs (shaded histogram) and unsuccessful pairs (open histogram), estimated from the date on which the male was first seen alone. Re-drawn from Williams (1973).

differences may extend back to the date of arrival in the breeding area since, in 1971, successful females arrived back on the Ythan significantly earlier than unsuccessful ones (Williams, 1973). There was, however, no significant difference in the dates of territory establishment between the two categories. Since older shelducks on average return earlier (section 3.1), Williams (1973) suggested that birds nesting earlier might be more successful because they were older. However, there is no consistent tendency for older Ythan shelducks to breed more successfully than younger ones (Makepeace & Patterson, 1980).

Coulson (1966) found that kittiwakes *Rissa tridactyla* which changed mate bred much less successfully than pairs which stayed together. However, Williams (1973) found no significant difference in hatching success between new shelduck pairs and those which had been together in the previous year.

Predation is unlikely to be a serious cause of nest failure in shelducks with their secure burrow nest sites. Only two of 39 nests found at the Ythan had signs of predation (table 6.3), possibly by a small mustelid *Mustela* (Young, 1964a; Patterson *et al.*, 1974). Jenkins *et al.* (1975) did not find any predation on 23 nests studied at Aberlady (table 6.3) and

Hori (1964a) states that at Sheppey 'Egg predators, other than human, were unknown and birds laid their eggs in nests in hay barns which contained the droppings of hedgehogs *Erinaceus europaeus* and of weasels *Mustela nivalis*.' Macdonald-Tyler (1956), however, found four shelduck nests apparently destroyed by foxes in Ireland. Human disturbance, including that by investigators, has been associated with some nest desertion by females (table 6.3) and Hori (1964a) reports nest-robbing and the 'milking' of nests by farm workers. One nest was also destroyed by a domestic dog.

Many failures of shelduck nests may be associated with disturbance by other shelducks. Young (1964a) found that his only two cases of multiple nests both failed, and Jenkins *et al.* (1975) suggested desertion as the cause of nine losses (47 per cent of those not due to human disturbance) at Aberlady. In at least four of these failures there was evidence that a second female had laid in the same nest. Disturbance or desertion occurred in seven out of ten nests with clutches of 12 or above but in only three out of 12 with smaller clutches. Their conclusion was 'that nest losses due to desertion were high, and that this was due in some cases to intra-specific disturbance'. Pienkowski & Evans (in press *b*) similarly found at Aberlady in 1976–9 that desertion occurred in 59 per cent of 22 multiple nests, but in only 27 per cent of 26 single ones ($\chi^2 = 4.09$, $P = 0.05$), and that some of the latter were associated with disturbance by rabbits and people. In the Camargue, I have seen many deserted multiple nests, some with over 30 eggs. Williams (1973) showed that the period of most nest failures coincided with the second seasonal peak of nest prospecting (figure 5.2) and suggested that the activities of these younger, non-territorial birds might cause some of the earlier breeders to desert. This could occur either through parasitic laying or through the incubating female's response to any disturbance of the burrow, discussed in section 6.4. However, the number of failures in any five-day period was not significantly correlated with the number of shelducks seen prospecting then and the overall hatching success each year was not correlated with the size of the non-territorial flock (Patterson *et al.*, in press *b*).

Although the information on possible interference by some shelducks in the nesting of others is sparse and not easy to interpret, there is nevertheless evidence of strong density-dependent interaction between the breeding pairs themselves. On the Ythan, the proportion of territorial pairs which were seen with broods declined significantly with increasing number of territories on the estuary (figure 6.7; Patterson &

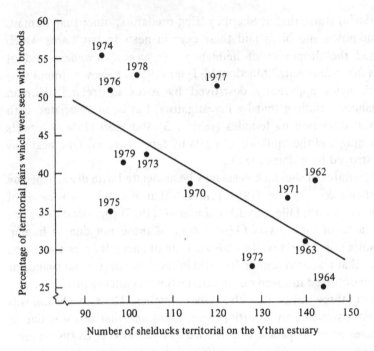

Figure 6.7 Nesting success of territorial pairs on the Ythan, in relation to the number of territories on the estuary each year (shown beside the points). The percentages are based on the number of pairs shown in table 6.4. The line is the calculated regression ($y = 80.42 - 0.34x$, $r = 0.678$, $p < 0.02$). From Patterson *et al.* (in press *b*).

Makepeace, 1979). The mechanism underlying this relationship is not known, but could be the degree of mutual interference between pairs. The significantly higher nesting success of the dominant pairs among nesting sub-groups is consistent with this, although dominants would be expected to gain an advantage in any competitive situation.

6.7 Individual strategies in nesting

How can a pair of shelducks maximise their chance of success-fully hatching a brood of ducklings? A first essential is to obtain a feeding territory, without which nesting seems not to be attempted. When prospecting, the pair have the options either of remaining alone or joining with others. In section 5.6 I discussed the possible costs and benefits of grouping, which seem to depend on the previous breeding success and dominance of the pair. Dominant and previously successful pairs would be best to remain on their own as much as possible, while subordinates and previously unsuccessful pairs might find some advantage in grouping with others.

In selecting a nest site, most females establish a nest of their own, but some opt to lay in the nests of other females. What are the relative advantages of these two strategies? Eggs laid in another clutch seem to hatch as successfully as those in single clutches, so long as the whole nest does not fail, but there are strong indications that such parasitised nests are more likely to be deserted. Thus it would seem to be best for a female to lay in a nest of her own if she can. However, if a female would not otherwise have a chance of breeding, e.g. if she did not have a territory or if she had been forced to abandon her own site just before or during laying, she might benefit by using another female's nest. Any of her eggs which were hatched and reared would represent a gain in fitness. Such parasitic females would also benefit by avoiding the time and energy costs of incubation, although such birds do not, as might be expected, make a second attempt to breed.

A female which detects that another has laid in her nest has the options of incubating the alien eggs with her own or deserting, since there is no evidence of any attempt at selective removal of strange eggs. It is perhaps surprising that so many females seem to desert in these circumstances, since there seems no cost to hatching success of the female's own eggs (table 6.2) (although the spread of hatching times may be increased). I shall also argue later than the presence of additional ducklings does not seem to affect fledging success (next chapter). Any increased energy costs in incubating a much larger clutch, and later in brooding more ducklings, should be offset to some extent by the ability of a bigger clutch or brood to keep themselves warm (Mertens, 1969). Possibly the chance of desertion of a parasitised brood is largely an anti-predator response to signs of disturbance of the nest. If a predator had indeed found the nest, the best strategy for the female might be to desert immediately, as I suggested earlier. It may not be possible for the female to identify the cause of disturbance and so distinguish between duck and predator activity. Presumably this response has been selected by predation in the past, since at present parasitic shelducks seem much commoner than predators of females on the nest.

6.8 The proportion of the population which breeds

The total breeding output of a local population of shelducks will be greatly affected by the extent of any non-breeding among the adults.

The most obvious possible non-breeders are the non-territorial birds. There is no evidence that shelducks without territories ever lay eggs (Young, 1964a; Williams, 1973; Patterson *et al.*, 1974). No Ythan male without a territory was seen consistently alone or later accompanying young. All females seen flying back to the estuary from their nests joined territorial males. About a quarter of the Ythan pairs seen with broods had not previously been seen on territories (mean 24.6 per cent, range 20.0–30.0, 1975–9). However, it was never possible to identify all the territory owners each year, especially on the upper parts of the estuary and on freshwater pools, where the birds swam for much of the time, making it difficult to identify rings. It is likely, therefore, that 'unknown' brood parents were unidentified territorial pairs, probably from freshwater pools. There is thus no substantial evidence that shelducks without territories ever breed.

Among territorial shelducks, early observations that broods were few relative to the number of adult pairs (e.g. Boase, 1951), and that some females seemed to be present on the territory throughout the season, led to suggestions that many of the territorial pairs did not breed. Young (1964a) suggested that some territorial pairs were never absent from the territory for long enough to have established a nest and some territorial males were never seen alone. He concluded from this that up to half of the territorial pairs did not lay in any given season. He associated this with gatherings in the nesting area, since most of the birds he saw in groups subsequently failed to produce young. This last finding, however, is not surprising since he made only one observation before 15 April and 70 per cent of his sightings of birds were in May and June, when young, non-territorial and already failed birds occur in the nesting area.

Young's (1964a) method of detecting breeding, by seeing the male alone on his territory, is usually the only one feasible, since it is usually impossible to find the nests of a large enough proportion of the population. The technique, however, is very sensitive to the frequency of observation. If clutches are lost during laying or soon afterwards, the short absences of the female would be detected only by frequent or prolonged observation and such pairs could easily be regarded as non-breeders. The method can be assessed by considering marked males which were later seen with ducklings, and finding what proportion of such successful pairs would have been detected before hatching by seeing the male alone on his territory. Since the females of these successful pairs would have been absent for much of the time over the

Table 6.5. *Percentage of male Ythan shelducks, later seen with broods, which were seen alone on territory at least once during the laying and incubation period of their mate. From Patterson et al. (1974) and data in Williams (1973)*

Year	Number of males	Percentage seen alone
1962	13	30.8
1963	10	60.0
1964	8	50.0
1966	11	100.0
1970	13	100.0
1971	14	100.0

Percentages in 1966–71 all significantly higher than in 1962–4 (χ^2 tests, $P < 0.05$).

whole laying and incubation period of 35–45 days they had a much higher chance of being recorded absent than had females which failed at some stage.

In 1962–4, just under half of the successful males were seen alone during laying and incubation (table 6.5 and Young, 1964a) while in subsequent years, with a higher frequency of observation on a sample of accessible territories, all the successful males were seen alone at least once. This shows that a low frequency of observation will fail to detect even some of the known successful pairs, although of course it does not show that even a high frequency of observation will detect early failures.

The proportion of territorial males seen alone in the whole Ythan population or in samples of the more easily observable territories was much higher in the later years of study than in 1962–4, when the frequency of observation was low (table 6.6). In 1971, when Williams (1973) observed a sample of territories four times each day, only one male out of 31 was never seen alone. The mate of this male had a noticeably bulging abdomen, which Young (1970b) found to be correlated with enlarged ovary and oviduct and the presence of eggs, so that she may well have laid. Thus, given a sufficiently high frequency of observation, there was no evidence of a substantial amount of non-breeding among the territorial Ythan shelducks.

Hori (1969) was able to find a large proportion of the nests in his study area and concluded that all or almost all of the territorial pairs on

Table 6.6. *Percentage of territorial male Ythan shelducks in which the male was seen alone on his territory at least once. From Patterson et al. (1974) and data in Williams (1973)*

Year	Number of males	Percentage seen alone
1962	71	62.0
1963	70	45.7
1964	72	43.1
1966	27	85.2
1970 ringed males	29	82.8
1970 other males	28	78.6
1971 ringed males	31	96.8

In 1966, 1970 and 1971 a sample of territories was observed intensively, once daily in 1966 and 1970 and four times daily in 1971. Percentages in 1966–71 all significantly higher than in 1962–4 (χ^2 test, $P < 0.05$)

Table 6.7. *Number of occupied nests in relation to the Sheppey breeding population. From Hori (1969)*

	1962	1963	1964	1965	1966	Mean
Estimated number of breeding pairs	70	73	65	115	147	94
Number of nests found	18	25	35	38	40	31
Number of nests identified[a]	61	64	60	100	125	82
Percentage of pairs with identified nests	87	88	92	87	85	87

[a] Nests were identified by females entering holes, hissing from holes, observation of flights by females back to the nest and observation of nesting groups.

Sheppey nested each year (table 6.7). Jenkins *et al.* (1975) also concluded that the number of nests found in their study area was 'much the same as the number of resident pairs and the presumption is that all of these attempted to breed'.

7

Care of the young

Parental behaviour normally involves a number of different activities: feeding the young, providing them with shelter and protecting them from predators and other dangers. By investing time and energy in this behaviour, even at some risk to their own chance of survival, the adults enhance their genetic fitness by increasing the survival of their progeny. Shelducks, like other parents of precocial young, do not feed their broods but merely accompany them while they feed themselves. The parents do, however, actively provide shelter and protection as I shall describe in this chapter.

7.1 Leaving the nest

Newly hatched young shelducks stay in the nest for some time, usually for at least 12 h and up to four days if there is a wide spread of hatching date within the clutch (Hori, 1964a). During this period, ducklings and female call frequently. Hori was able to approach very close to a nest inside a shed and described a monosyllabic 'aarrk' and a soft running 'ugg ugg ugg' given continuously for long periods by the female. I installed a microphone in a nest burrow over the hatching period and heard similar calls, along with piping responses from the ducklings. All of the calls are given very softly and can only be heard at very close range. It is likely that the ducklings become imprinted on the mother's voice while in the nest (Gottlieb, 1965). Since the ducklings' first hours are spent in the dark, it is obviously functional to use auditory rather than visual characteristics of the mother as a basis for imprinting.

During the hatching period the female continues to leave the nest to feed, and Knight (1925) found ducklings covered by down while their mother was absent. Most of the Ythan females whose activities were monitored by watching at the territory, or by recorders in the burrow, spent only very short periods off the nest during the days of hatching. A few indeed seemed not to leave at all in the last 12 h before the ducklings left the nest. This change in the routine established over the 30 days of incubation may be the stimulus which leads to the male remaining close to the nest as hatching approaches. However, there may be other changes in the female's behaviour to which the male responds.

The ducklings are led from the nest at any time of day, although most commonly early in the morning (Hori, 1964a). No overnight departures have been reported, so that ducklings reaching the critical stage in the late evening or night will have to wait until morning, creating a peak of departures then. Young (1964a) suggested that the ducklings' arrival on the shore usually coincided with a rising tide, but there is no general evidence for this. One marked Ythan female, seen arriving with her brood in two consecutive years, reached the estuary at different times of day and at different stages of tide in the two years.

The male is usually present when the ducklings first leave the nest. At two nests at the Ythan where the emergence of the brood was filmed (Royal Society for the Protection of Birds film *Shelduck*) the females left the nests soon after dawn and returned quickly with their mates, which landed and stayed at the nest entrance. Again, it is not known what stimulates this change from the male's usual avoidance of the nest itself, although the obvious excitement and agitation shown by the females is likely to be important. At one of the filmed nests the male appeared to be alarmed by the presence of the cameraman and flew off. The female repeatedly left the nest and returned with him, only to have him leave again while she was inside the nest hole. Finally she brought out the ducklings without him, as Hori (1969) saw several times on Sheppey.

The departure of the young from the nest hole is preceded by an increase in frequency and loudness of the female's calling, usually the long 'ak ak ak . . .' note (Hori, 1969). The duck may call inside the

Figure 7.1 Newly-hatched shelducklings. Photo by M. Makepeace.

burrow and lead the ducklings out, or may call from outside until they approach her, as happened at one of the filmed Ythan nests. Females also call outside tree holes until the young climb out and jump down to the ground (Hori, 1964a,c, 1969). In some Sheppey tree nests, the young fell into the bases of the hollow trunks but were often able to climb out (Hori, 1969). Rittinghaus (1956) has shown how the well-developed legs and claws of young shelducks enable them to jump and cling effectively. Some young, however, were trapped in hollow trees for several days and would die unless rescued. Hori (1964c, 1969) considered that some ducklings escaped from hollow trunks so deep and smooth sided that they must have been carried out by the female. He reports that this had been seen by farm workers but there are no first-hand accounts of females carrying young.

Once they emerge, the ducklings are led quickly to the shore, usually by both parents. This may involve an overland journey of several kilometres and the crossing of obstacles such as walls, roads and occasionally built-up areas (Nelson, 1944). The journey seems to be

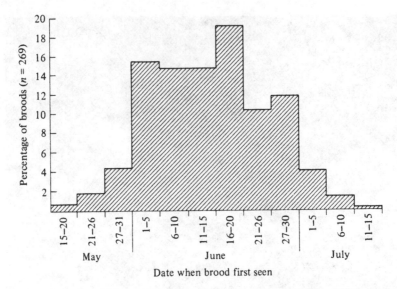

Figure 7.2 Timing of emergence of broods on the Ythan, 1970–9. Data for 1970–2 from Williams (1973).

made quickly and without a break, since few families are seen in transit and most are first seen already on the water.

Most broods appear on their feeding areas in June (figure 7.2). The earliest brood seen on the Ythan was a pair with three day-old ducklings on 18 May 1977, although a pair seen with a single duckling three to five days old on 20 May 1978 had probably hatched it earlier. Both of these were small broods which, if they resulted from small clutches, would have had an early start to incubation. Hori (1964a, 1969) saw broods on Sheppey on 28 May and Gillham & Homes (1950) saw some as early as 22 May elsewhere in Kent. South & Butler (1955) reported a brood of six from inland Berkshire on 30 May 1954 and Staton (1945) saw a brood of ten in Nottinghamshire on 29 May. The northern Ythan population was thus slightly earlier than those further south in Britain. The date of brood appearance seems similar in Europe. Ducklings were reported by 18 May in Russia (Dementiev & Gladkov, 1952), by 27 May in Denmark (Lind, 1957) and by the end of May in Belgium (Maebe & van der Vloet, 1952). Enkelaar & Lebret (1966) showed a distribution of emergence dates for shelduck broods in the Netherlands similar to that for the Ythan. Their earliest brood was estimated, from body size and plumage development when first seen, to have emerged during 15–20 May 1963 and their latest brood hatched during the first five days of

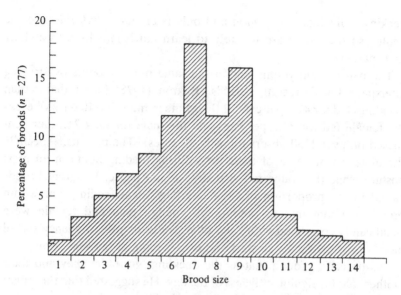

Figure 7.3 Number of ducklings in broods when first seen on the Ythan, 1962–79. Data from 1962–64 from Young (1964*a*).

August 1964. They showed a peak of hatching in mid-May in 1963 and 1964. The latest brood on the Ythan was first seen on 15 July 1970 (Williams, 1973).

Brood size on first emergence on the Ythan varied from one to 22 ducklings, with a mode of seven and a mean of 7.3 ± 0.2 (figure 7.3). Most broods (58.1 per cent of 277) had six to nine ducklings and only 4.3 per cent of broods had more than 12 young. Hori (1969) found mean brood sizes in different years (1962–6) of 6.8 to 9.6 ducklings but in his area on Sheppey up to 27 per cent of the broods had more than 12 ducklings. The higher proportion of larger broods on Sheppey than on the Ythan probably resulted from the difference in frequency of multiple nests in the two areas (section 6.3) but could also be due to some amalgamation of broods before being seen, as I will discuss later.

7.2 Feeding

The ducklings' yolk reserves maintain them for a few days (Kear, 1965) while they learn what is edible and how to feed efficiently enough to start gaining energy. The parents can presumably contribute to this process by taking the ducklings quickly to areas of high food abundance and by keeping them there, but do not otherwise appear to help the young to learn to feed. In any case, the ducklings at first feed mainly by pecking individual *Corophium* and only later sieve *Hydrobia* like the adults, so the young are unlikely to learn much initially by watching their parents.

The parents and young feed at the same time, in bouts of foraging interspersed with resting periods. Buxton (1975) found that Ythan ducklings fed for 41.7 per cent of their time in their first 10 days whereas the female fed for 33.5 per cent and the male for only 21.8 per cent (based on over 1260 observations in all cases). The parents, especially the male, spent some of their time during feeding bouts in an alert posture while the young fed virtually without pause. Different broods varied in the proportion of their day they spent feeding, but within broods the amounts of feeding by young, male and female were significantly correlated with each other, reflecting the tendency for all members of the family to start and stop feeding at the same time.

Buxton (1975) found that as the ducklings got older they and their mothers fed for significantly less of the day. He suggested that this might be due to more efficient feeding by the ducklings as they began sieving like the adults and to decreased energy losses as their feathers grew and their surface-to-volume ratio decreased. Females might also be expected to feed more immediately after incubation, to make up any loss of weight. The males did not show a similar decrease in the amount of time they spent feeding as the ducklings got older.

7.3 Parental behaviour
Brooding

Between bouts of feeding the female broods the young, at least while they are small. In 28 Ythan broods, observed every ten minutes for three-hour periods, ducklings in their first week were brooded for 9.6 per cent of the day whereas by three weeks the amount had dropped to 1.6 per cent (figure 7.4). It is not clear whether females or ducklings initiate spells of resting and brooding. Usually on the Ythan females seemed to be first to stop feeding and walk to the high tide line followed by the young, but sometimes ducklings were seen to approach a female first and so perhaps stimulated her to brood. Redhead (1979) found that females brooded small young more when the weather was more chilling (low temperature and/or high wind speed). Conversely, on warmer days, ducklings were more likely to rest alone or in small groups without going under the female (figure 7.5). The total amount of time spent resting did not change with weather, only the proportion of it which was spent brooding.

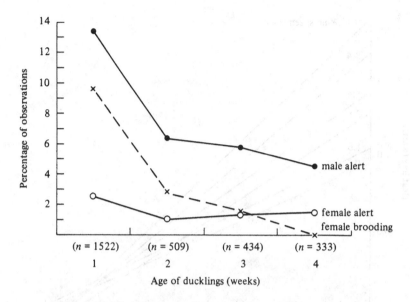

Figure 7.4 Percentage of time spent alert and brooding by parent shelducks with ducklings of different ages.

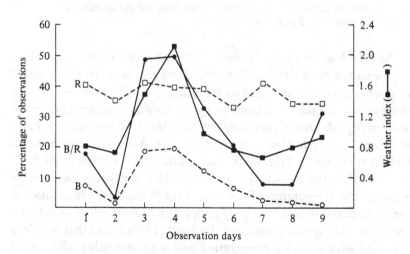

Figure 7.5 Changes in the percentage of time spent resting and brooding, in relation to an index of chilling weather. The letters indicate different types of behaviour: R, resting but not being brooded; B, being brooded; B/R, the percentage of resting time in which the ducklings were being brooded. From Redhead (1979).

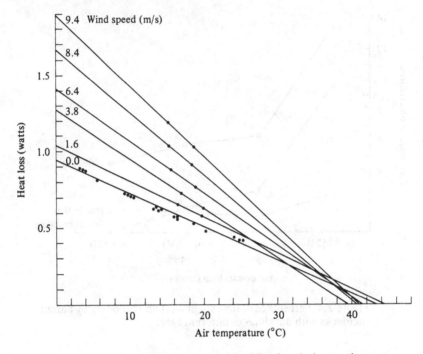

Figure 7.6 Heat loss by the model duckling in relation to air
temperature and wind speed (shown on the lines). The points shown
are the actual measurements and the lines are the calculated
regressions. From Redhead (1979).

The heat loss by young shelduck under different weather conditions
was measured using an 'electric duckling', which I made from the skin of
a day-old duckling found dead on the Ythan. A small three watt heating
element was encased in a block of epoxy resin and inserted inside the
body cavity. A circuit (designed by Jim Bruce of the Scottish Farm
Buildings Research Unit and built by Jim Anderson of Aberdeen
University) controlled the temperature of the block within 1 °C of 40 °C
and recorded the amount of time for which the heater was on to
maintain this temperature. From this, I could calculate the amount of
energy required to maintain body temperature under different external
conditions. Using this equipment Redhead (1979) found that heat loss
increased with lowering temperature and with increasing wind speed
(figure 7.6).

In a high wind speed, heat loss was less when the duckling was behind
a low shelter (little more than its own height when sitting) suggesting
that even moderately long grass might provide shelter. Since the

duckling's down had lost its normal waterproofing it was not possible to test the effect of rainfall. In some preliminary trials I found that net heat loss was greatly reduced or even reversed by heat gain from direct radiation in sunlight. These results from the 'electric duckling' thus explained why ducklings are brooded more in chilling weather when their rate of heat loss should increase. Mendenhall (1979) made similar field observations on brooding by eiders and also found that the amount of brooding was higher in bad weather and decreased with the age of the ducklings. Eider ducklings also did not change the amount of total resting time with weather and indeed fed more in bad weather, presumably to offset their greater energy losses.

Anti-predator behaviour

Shelduck parents spend a lot of their time alert. The males, especially while foraging, are particularly vigilant (figure 7.4). Potential predators which come within a few hundred metres of the brood are frequently subjected to a flying attack and pursuit by the male. Gulls, especially the great black-backed *Larus marinus*, herring *L. argentatus* and lesser black-backed *L. fuscus* gulls, crows *Corvus corone* and *C. cornix* and herons *Ardea cinerea* (T. A. W. Davis, 1975) are particularly likely to be attacked. However, other species such as black-headed gulls *Larus ridibundus*, common terns *Sterna hirundo* (Young, 1964a) and waders, which are unlikely to prey on ducklings, are also attacked. On the Ythan in 1972, 72 per cent of 92 attacks were on gulls, mainly black-headed and herring, and 11 per cent were on waders (Makepeace & Patterson, 1980). J. Davis (1975) described how male shelducks drove off great black-backed gulls while the female kept the brood together, in contrast to mallard *Anas platyrhynchos*, in the same situation, where the female alone was not able to protect the young adequately.

Shelducks do not attack large mammalian predators, which are likely to be more dangerous to the adults. Instead, when a brood is approached by people or domestic dogs the adults fly round in wide circles giving alarm calls ('tolling', Sowls, 1955; Hori, 1964c). The ducklings respond by rushing to the nearest water, where they dive and scatter if the predator comes close, or to cover where they hide.

If a large mammalian predator comes close to the ducklings, the adults may intensify their behaviour into a distraction display. This begins by the birds flying low and very slowly away from the predator. The legs are often dangled and may splash along the surface of water. Occasionally an adult may show violent 'injury-feigning' similar to that

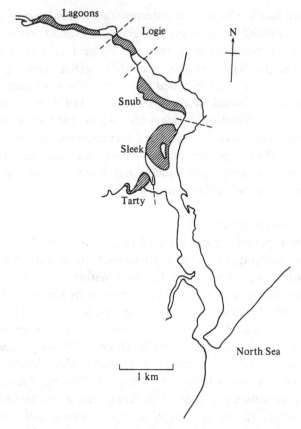

Figure 7.7 Main duckling nursery areas on the Ythan. Re-drawn from Makepeace & Patterson (1980).

shown by mallard (Hori, 1964c). Young (1964a) described how a female led her brood away from two men while the male distracted their dog and led it away in the opposite direction. The birds' behaviour can change quickly if the predator situation alters. T. A. W. Davis (1975) found that a male which had just attacked and driven off a heron showed a distraction flight to the observer immediately afterwards.

7.4 Brood ranges

Surprisingly, in spite of the effort expended by the male shelduck in defending a territory during the nesting period, the ducklings are rarely taken there (Hori, 1964a, 1969; Williams, 1973). Instead the territory is abandoned and the family settles at a new site, in the 'nursery areas' where most of the broods go. On the Ythan the main

Table 7.1. *Fidelity to nursery areas on the Ythan. Birds which had at least three broods in 1972–7. From Makepeace & Patterson (1980)*

	Number of pairs	Percentage which used same area in next year
Pairs which stayed together	20	50
Male with a new mate	3	0
Female with a new mate	7	29

None of the differences between categories is statistically significant.

nursery areas were either wide mudflats with undisturbed shores (figure 7.7) or secluded tributary streams and freshwater pools with abundant cover on the banks. Other parts of the estuary, used for territories (figure 4.5) but not by broods, tended to be narrower and close to roads or other sources of disturbance. The main brood areas were concentrated further upstream than were the territories. On Sheppey, Hori (1969) similarly found that the main nursery area was a large creek which was fairly inaccessible to people, whereas most of the territories were on narrow channels in the grazing marshes.

Although, in general, the brood is not taken to the former territory, Williams (1973) found that of 21 Ythan pairs whose former territories had been in sections used later as nursery areas, 57 per cent settled near the old territory site. However, the same boundaries were not observed by the brood.

Shelducks usually take their broods to the same general area in successive years. Only half of the Ythan pairs which stayed together and hatched broods in successive years went back to the same section of the estuary (table 7.1) but 90 per cent used only one or two sections over three to six years, out of the six sections available. Some pairs were seen to return first to their previous nursery area but were evicted by another brood already there and moved to a second area, usually adjacent. Williams (1973) recorded similar cases. If there was a change of mate between years, fewer birds used the same nursery area as in the previous year (table 7.1). There seems to be no tendency for families of shelduck to have traditional nursery areas. Of 23 Ythan-reared females rearing broods for the first time, 39 per cent returned to the nursery area where they had been reared. This percentage is not significantly different from the expectation if birds went randomly to the nursery areas available (χ^2 test).

Table 7.2. *Sizes of brood ranges (ha) on the Ythan in relation to the age of the ducklings. From Williams (1973)*

| Brood | Plumage class | | | |
	1	2	3	4
1	5.0	6.0	8.0	15.5
2	6.0	10.0	19.5[a]	—
3	6.0	5.0	5.0[a]	—
4	13.0[a]	17.0[a]	—	—
5	8.5[a]	5.0[a]	—	—
6	19.0[a]	15.0[a]	9.0[a]	—
7	6.0[a]	12.0[a]	—	—
8	—	5.5	9.0	55.0[a]
9	12.0	14.0[a]	22.5	16.5[a]
Mean	9.4 ± 1.6	9.9 ± 1.5	10.8 ± 1.4	29.0 ± 10.6

[a] These ranges were probably underestimated.

Broods on the Ythan usually stayed within the same section of the estuary throughout the ducklings' development, but eight per cent of 75 broods in 1970–2 moved from one area to another (Williams, 1973). Two of the movements were from freshwater pools to the Ythan estuary, involving distances of 2–3 km and the loss of several ducklings.

The sizes of ranges occupied by broods on the Ythan were estimated by Williams (1973) by plotting the position of each brood once each day. Some broods were observed continuously through whole tidal cycles and their positions were plotted every five minutes on a 1 ha grid. The data were analysed using the area–observation curve and boundary strip method already used for estimating territory size (section 4.6). Range size averaged 9.4 ha for broods in their first week and generally increased with the age of the ducklings (table 7.2). There was also, however, a tendency for brood ranges on the Ythan in 1975 and 1976 to decrease as more broods settled in each area (Makepeace & Patterson, 1980). Buxton (1975) found that ranges were usually smallest in areas with the highest density and biomass of *Corophium* and *Nereis*, excluding one area with very small ranges where the birds were probably feeding on other organisms.

7.5 Aggressive interactions between brood pairs

Shelducks' brood ranges are not exclusive to each family but overlap widely between adjacent broods, particularly when several are

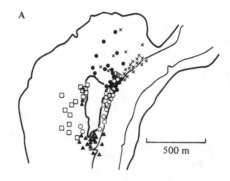

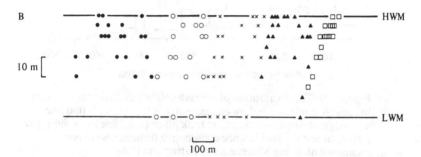

Figure 7.8 Positions of adjacent shelduck broods on the Ythan over three-hour periods, in an area with overlapping ranges (A) and without (B). In area B, a section of straight narrow shore, the vertical axis has been exaggerated for clarity.

crowded into one area (figure 7.8). However, within the overlapping areas the broods appear to avoid each other, and on the Ythan they usually stayed 100–300 m apart (figure 7.9). Ladhams (1971) described how two pairs on Chew Valley Lake, Somerset both moved all over a small pool. The young of the two families intermingled freely although 'rarely entering each others territory'. On the Ythan, the first broods to emerge seemed to occupy the more remote areas of the estuary, the Lagoons, Logie and Tarty (figure 7.7), but they were usually joined later by other broods and were not able to keep the areas for their exclusive use. An exception to this was Tarty, which was so narrow that one pair could often prevent others from entering.

When pairs with broods come close to each other there is commonly aggressive interaction between the two sets of parents, especially when one of the families has newly settled in an area already occupied by

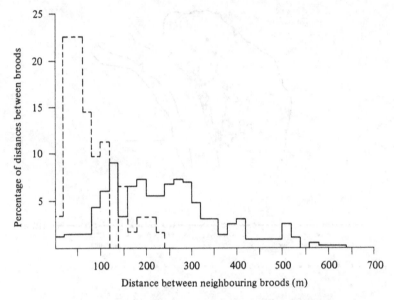

Figure 7.9 The distribution of nearest-neighbour distances between broods, five minutes before an aggressive interaction (dotted line, $n = 62$) and when no interaction took place (solid line, $n = 398$). No statistical test is possible since consecutive distances were not independent. From Makepeace & Patterson (1980).

another. Hori (1964b) also described conflict and severe fighting between brood parents, 'especially later in the season when overcrowding occurred . . .', and an incident was reported by Roseveare (1951). On the Ythan, most interactions occurred when broods were less than 100 m apart (figure 7.9).

The first sign of aggression was hunching of the back in the Head-down posture, adopted mainly by the males, where the head was lowered and the feathers of the back raised. This was commonly followed by Head-throwing by the males and Inciting by the females as the two broods converged. One of the broods might then move away and hostilities would die down. However, if the broods got closer, the interaction could escalate into one where the adults made attack charges without any actual contact, or one where attacks and fights with contact occurred. Males nearly always attacked and fought the male of the other brood, and the most violent fights occurred with all four birds of two pairs fighting, male with male and female with female. Birds were pecked (often with feathers pulled out), beaten by wings, and often pushed under water or held down on the ground, but no serious injuries were recorded. Some possibly redirected attack was seen when brood adults suddenly left the interaction for a few minutes and attacked a nearby bird, not connected with the interaction.

The outcome of some interactions was clear cut; the winner stood his ground, and the loser retreated. Occasionally, however, both broods moved apart at the same time, especially when the broods had already had frequent encounters with each other. Both brood parents appeared to be necessary for success in these interactions. A female alone was never seen to win against an attacking male, and so could not defend her ducklings. A male alone could win fights but seemed to be unable to keep the ducklings with him (the female did all the calling together of ducklings).

Attacks are also made on other waterfowl which come close to the ducklings. On the Ythan, 11 per cent of 92 attacks on other species were against eiders and mallard. Ladhams (1971) also found that coots *Fulica atra* and all the common duck species were attacked, although mute swans *Cygnus olor* were not.

Attacks are often made upon ducklings by the parents of another brood. Interactions on the Ythan frequently started by ducklings being attacked by adult shelducks who were in turn attacked by the parents defending their ducklings. Attacks on ducklings by adults from another brood could be severe. Ducklings were run down and bitten, picked up and beaten on the ground, or repeatedly pushed under water where they usually dived to escape. No duckling was seen to be killed, but some were badly beaten. Such attacks occurred in 21–24 per cent of all interactions (1975 and 1976). Hori (1964*a*) also recorded an attack on the ducklings of one brood by the parents of another and Ladhams (1971) saw the attacking parents dive underwater, apparently in pursuit of diving ducklings of the other brood.

The distance between the ducklings and their parents may change during an interaction. On the Ythan, in the brood being attacked there was a slight increase in the mean distance for older ducklings (figure 7.10) but none for younger ones. In some cases, the ducklings ran together in a tight group close to their parents after being physically attacked by other adults. In the attacking brood, especially with older ducklings, there was a considerable increase in the mean distance between the young and their parents during an interaction (figure 7.10), largely through the adults leaving their brood to make the attack. Occasionally, single ducklings became separated from the rest of the brood and had to run or swim over 100 m to re-join it. This increase in distance from

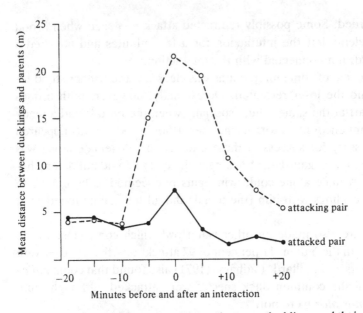

Figure 7.10 The mean distance between ducklings and their parents, before, during and after aggressive interactions between brood pairs. Re-drawn from Makepeace & Patterson (1980).

the parents might increase the risk from predators and one duckling was seen taken by a gull during an interaction.

The frequency of aggressive interaction

The rate at which aggressive interactions occurred between brood parents was measured on the Ythan by watching adjacent broods for prolonged periods (Makepeace & Patterson, 1980). In areas of the estuary occupied by several broods the frequency of aggressive interaction increased significantly with increasing density of broods in both 1975 and 1976 (table 7.3). The frequency of interactions which involved physical attack on the ducklings of either of the broods also increased with brood density in both years although the correlation was barely significant. Partial correlation, controlling for seasonal changes in density and interaction frequency, reduced the correlations to some extent, particularly the one between density and physical attacks on ducklings in 1976.

Experimental manipulation of brood density

In an area of the Ythan estuary containing four or five broods, the density was increased experimentally by a person moving upstream

Table 7.3. *Correlations and partial correlations between the frequency of aggressive interactions among broods and the density of broods per hectare in the observation area. From Makepeace & Patterson (1980)*

	Total interactions per brood per hour			Physical attacks on ducklings per duckling per hour		
	Kendal τ	DF	P	Kendal τ	DF	P
1975 Kendal correlation coefficient	+0.51	16	0.004	+0.33	16	0.053
Partial correlation, controlling for date	+0.40	15	—	+0.27	15	—
1976 Kendal correlation coefficient	+0.42	20	0.004	+0.29	20	0.050
Partial correlation, controlling for date	+0.35	19	—	+0.17	19	—

Table 7.4. *Mean frequency of aggressive interaction between broods during crowding experiments. The samples* (n) *are one-hour observation periods. From Makepeace & Patterson (1980)*

	Control periods	Experimental periods
Mean interactions per brood per hour	0.23 ± 0.10	1.30 ± 0.17
	$n = 8$	$n = 5$
	Mann-Whitney U $= 1, p < 0.002$	

to push the nearest broods closer together for three hours. The experimental session was preceded and followed by undisturbed control sessions on adjacent days with the same total number of broods in the area and with similar weather. Since only the broods nearest to the disturbance were likely to be affected, the measurement of density over the whole area would be misleading and was replaced by estimates of mean nearest-neighbour distance between broods, measured from plots of the position of each brood every five minutes. All aggressive interactions between the broods were recorded as before. Mean values for nearest-neighbour distance and frequency of interaction were calculated for each hour of observations.

The presence of a person at one end of the area did produce an increase in density; the mean nearest-neighbour distance of the two broods nearest to the disturbance was significantly reduced relative to the mean distance between broods without disturbance (Makepeace & Patterson, 1980).

The frequency of aggressive interaction between the artificially crowded broods was significantly higher than that between control broods (table 7.4). The frequency of all aggressive interactions in a given hour of the experiment also increased significantly with decreasing mean nearest-neighbour distance between the broods ($r = 0.558$, $n = 59$, $P < 0.001$).

Thus the experimental crowding of broods on the Ythan showed clearly that aggressive interaction between broods did increase with increasing brood density, and so confirmed the earlier observations: conflict between shelduck broods does seem to increase with increasing number of broods crowding into a nursery area.

7.6 Brood-mixing and creches

Although some shelduck broods remain as discrete family units until the young fledge, mixing of broods to form creches containing ducklings of varying ages has been widely reported (Kirkman, 1913; Boase, 1938, 1951, 1965; Coombes, 1950; Gillham & Homes, 1950; Dementiev & Gladkov, 1952; Bannerman & Lodge, 1957; Hori, 1964a; Williams, 1974). The same phenomenon occurs in other waterfowl and 11 species in which it has been described were listed by Williams (1974). In shelducks the mixed broods can range in size from only a few ducklings to creches of over 100 (Boase, 1951; Hori, 1964a) but most mixed broods on the Ythan contained fewer than 20 ducklings (Williams, 1973).

The mixing of broods can occur by the ducklings from different families becoming intermingled while feeding (Ladhams, 1971). Williams (1974) found that, especially on large mudflats, even small ducklings tended to range ahead of their parents and could be up to 50 m away from them. It was possible for the ducklings of two broods to mix while the parents were still 100 m apart. When the adults called, the ducklings did not necessarily return to the correct parents but often followed other ducklings to the nearest adult. On three occasions Williams (1974) saw this result in one pair collecting all the young of two broods, while in another six cases only one or two ducklings were transferred between broods. Young (1964a) considered that most mixing of broods occurred during feeding.

Mixing can also occur during aggressive interactions between brood parents. Especially if these are prolonged, the ducklings of both broods can wander off and mix together. After the interaction, one pair may then collect some or all of the other pair's ducklings. Williams (1974) saw five interactions which led to mixing, although many broods remained intact after conflicts. Hori (1964a) considered that attacks on ducklings by adults, particularly by the pair tending them (see below), were the principal cause of creching on Sheppey.

The readiness of young shelducks to join other broods is puzzling since it suggests a lack of imprinting on the individual characteristics of the mother, even her voice which the young apparently respond to when leaving the nest. The ducklings appear to be imprinted on the general specific characters of shelducks and perhaps also on other ducklings. I have seen no tendency to follow other species of ducks, e.g. eider and mallard, even when these pass close by with ducklings. For a young shelduck, which is likely to range far from its parents and so risk being

Table 7.5. *Percentage of broods of different ages involved in brood-mixing, either as donors or recipients. From Williams (1974)*

| | Plumage class | | | |
	1	2	3	4
Percentage involved in mixing	52.3	43.2	17.8	4.5
Number of broods	65	37	28	22

separated from them, there may well be an advantage in approaching any parent shelduck and attempting to be adopted. Older ducklings may discriminate their own parents, perhaps by voice. I have twice seen mixed broods of large ducklings separate correctly when one of the females called and her own young left the creche to join her. I will discuss later the discrimination of young by the parents.

Age of ducklings which mix

Most mixing of broods occurs while the ducklings are small. Young (1964a) found that in six out of ten mixed broods on the Ythan in 1962 and 1963, the ducklings were all under 12 days of age; in the remainder they were all under 18 days. Williams (1974) similarly found more brood-mixing among ducklings in their first weeks of life than among older ducklings (table 7.5). This was partly because broods of young ducklings were commoner, the number of older broods being reduced by brood amalgamation and mortality. Even allowing for this, however, mixing was proportionally more common among the younger ducklings.

Mixing also tends to occur between broods in which the ducklings are similar in age. Williams (1974) found that almost all transfers of ducklings were between broods whose ducklings were in the same age class or adjacent age classes (table 7.6). Again this may be because there was a strong tendency for most of the ducklings in an area to be similar in age, but the adults may also have discriminated against young different in age to their own (see below).

The frequency of brood-mixing

The amount of mixing between broods varies considerably between areas, between years and within the season. Hori (1964a) found on Sheppey that up to the first week of July only 29–52 per cent of the ducklings in the area were in mixed broods whereas after mid-July

Table 7.6. *Number of transfers and number of ducklings transferred (in brackets) between broods of different ages. From Williams (1974)*

		Plumage class of recipient brood				
		1	2	3	4	Total
Plumage	1	22 (66)	4 (10)	—	—	26 (76)
class of	2	6 (17)	8 (9)	1 (1)	—	15 (27)
donor	3	—	—	5 (7)	1 (1)	6 (8)
brood	4	—	—	—	—	0
	Total	28 (83)	12 (19)	6 (8)	1 (1)	47 (111)

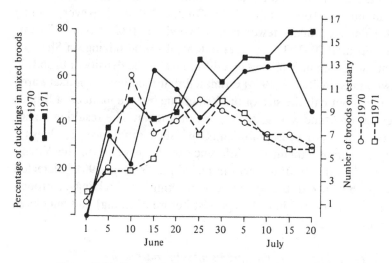

Figure 7.11 Seasonal changes in the percentage of ducklings in mixed broods (closed symbols) and the number of broods present simultaneously on the Ythan (open symbols) in 1970 and 1971. Re-drawn from Williams (1973).

up to 92 per cent were mixed. The transition took place within a week or so in early July which coincided with the onset of the moult migration. Williams (1973) reported that the amount of mixing increased during the duckling season as a larger number of broods emerged on the estuary (figure 7.11). On Sheppey crowding increased during the season and it seems likely that the density of broods in an area and the frequency of interaction between them are important causes of brood-mixing. Young (1964a) found most brood-mixing in the most crowded parts of the Ythan.

Table 7.7. *Extent of brood-mixing in different areas. From data in Hori (1964a), Young (1964a) and Williams (1974)*

	Sheppey[a]			Ythan					
	1961	1962	1963	1962	1963	1964	1970	1971	1972
Percentage mixing	82	91	92	14.8	27.3	0	71.5	63.0	76.4
Number of ducklings	131	69	199						
Number of broods				27	22	18	21	27	17

[a] Seasonal peak figures.

By late July in 1961–3 on Sheppey, almost all of the ducklings present were in mixed broods (table 7.7). On the Ythan, however, Young (1964a) detected many fewer mixed broods in 1962–4 than Williams (1974) did in 1970–2. The greater extent of brood-mixing on Sheppey than at the Ythan may be due to a difference in the density of broods in the two areas. Williams (1974) estimated that the Sheppey nursery area was only about half the size of the Ythan one but the number of broods present at one time was up to 50 per cent higher. The resulting higher density could well cause more mixing. Such differences in density, however, cannot explain the difference between years on the Ythan, which may be due to differences in technique. Young (1964a) detected many of his mixed broods by dye-marking of ducklings (through injection of food dyes into the egg just before hatching) and may have missed other cases.

The behaviour of the parents after brood-mixing

When two shelduck broods become mixed the accompanying adults commonly become aggressive towards some or all of the ducklings in their care. Williams (1974) found that especially (though not exclusively) when the ducklings varied in size, one or both of the adults began to chase the young. Any that were caught were pecked, often vigorously, on the back, neck and head. When chased on the water, some ducklings were forcibly immersed. Williams (1974), however, did not see any ducklings drowned or killed by such attacks. Where the ducklings differed markedly in age and size the adults singled out the alien ones and attacked them exclusively. I have seen males consistently attacking strange young which had mixed with their own smaller or larger ducklings. Young (1964a) recorded attacks on ducklings known to be alien by their dye marks, although it is possible in these cases that the dye enabled the adults as well as the observer to distinguish the young. Where the ducklings varied only a little in size, Williams (1974) found that all members of the brood, including the adults' own young, were attacked. I have made similar observations and it seems that in these cases the adults can detect that strange young are present but cannot reliably identify them, especially once the attacks begin. In most cases of mixing where the ducklings are all the same age and size the accompanying adults show no obvious aggressive response and no sign of having recognised the presence of strange young. There thus seems to be no individual recognition of young, at least when they are small, since the adults only discriminate against strangers when they differ noticeably in size.

Most attacks by adults on ducklings in their care last for only a few minutes before the birds resume normal parental behaviour. Williams (1974), however, saw one pair which attacked their brood for 30 min and the male then continued spasmodic attacks for a further half hour. I saw another extreme example in 1975 where all of a creche of 20 ducklings, six of whom had been acquired the day before, were hounded for 3.5 h by the accompanying adult female. The male of the pair tried to defend the ducklings, and frequently attacked his own female when she attacked a duckling (Makepeace & Patterson, 1980). In most cases, however, once attacks had ceased they were not resumed, even in broods watched for up to five hours after the incident (Williams, 1974). None of the attacks seen on the Ythan appeared to drive ducklings away from the adults (Williams, 1974; Makepeace & Patterson, 1980).

All the attacks on ducklings by the accompanying adults seen by Williams (1974) followed the transfer of strange ducklings into the brood and. Young (1964a) also found the two events to be closely associated. On the Ythan in 1975 and 1976, adults attacked ducklings in their care after 10.6 and 4.0 per cent of aggressive interactions between brood adults, almost all involving known exchange of ducklings (Makepeace & Patterson, 1980). However, in 1977, one incident was recorded where there was no question of the brood having been involved in creching. A new brood repeatedly approached an established brood then faltered, and both the male and female started pecking their own ducklings for a few minutes before retreating. It is possible that aggression was redirected to the ducklings because the parents were afraid to attack the other pair of adults (Makepeace & Patterson, 1980). Apart from this one incident, however, all the workers on the Ythan have found that attacks on ducklings follow brood-mixing and can be interpreted as the adults attempting, unsuccessfully, to drive away ducklings which they recognise as not their own.

Hori (1964*a*, 1964*c*, 1969), however, while recording similar behaviour, had a very different interpretation. He considered that the ducklings being attacked were the adults' own young and concluded that the parents were attempting to drive away their ducklings so as to be free to migrate earlier and in better condition. Coombes (1950) had also earlier interpreted creching as a device to 'release' adults for the moult migration. However, none of the adults or ducklings involved was described as being marked and Hori does not discuss the possibility of prior mixing of young even though this was very common on Sheppey. One case occurred on 2 July (Hori, 1964*a*) when there were few creches, but even then mixing cannot be excluded. One incident occurred immediately after an aggressive interaction with another pair; thus, the possible mixing of young before attacks on the ducklings cannot be excluded in Hori's (1964*c*) observations at Sheppey, whereas the association between the two events has been confirmed repeatedly at the Ythan by direct observation and by marking of both adults and ducklings. However, it is possible that there are differences in behaviour between the areas. For example, the high density of broods at Sheppey might stimulate redirected attacks on ducklings, as I suggested for an isolated incident on the Ythan.

Which shelducks accompany creches?

Hori (1964*a,c*) considered originally that failed breeders (birds which had lost their nests) took over broods abandoned by their parents. However, later he found by marking females that four seen with creches had all hatched their own broods earlier in the season (Hori, 1969). Williams (1974) found that all of the many mixed broods on the Ythan were accompanied by pairs which were known to have hatched ducklings. He never saw failed breeders or non-breeders accompanying or attempting to accompany any ducklings. Nevertheless, I have seen non-territorial shelducks approaching or following broods with apparent interest in the ducklings and Hori (1964*c*) has made many similar observations. Such birds are usually driven off quickly by the brood male.

Williams (1974) found no tendency for particular adults consistently to accumulate ducklings in different years. Of 15 females which hatched young on the Ythan in two consecutive years, four had mixed broods in both years, five had in one year but not the other and the other six did not have mixed broods in either year. Males showed a similar distribution.

Pairs which lost all of their young to another brood tended to remain close to the mixed brood for some time. Williams (1974) saw two incidents where mixing was the result of aggressive interaction and where one set of parents was driven from the immediate area by the other male. In four other cases the pair which lost their young stayed in the vicinity of the mixed brood for about a day. Within two days, however, all six pairs had joined the non-territorial flock and were not seen with young for the rest of the season.

8

Duckling survival

Being mobile while still very small, precocial young birds are very vulnerable to a number of hazards, the principal ones being the risk of predation and the danger of chilling when not being brooded by a parent. As a result, most suffer a high mortality in their first week or ten days of life (Williams, 1974). Many may also have to face competition for space and resources from other families, and in some species aggressive interactions may result. As I discussed in the previous chapter, shelducks often exchange ducklings between broods and may even form large creches. In this chapter I will discuss the effect of this behaviour and of other hazards on the survival of the ducklings.

8.1 Survival rates

A large proportion of shelducklings die before fledging. Boase (1951) suggested that about one third of the ducklings on the Tay were lost, and Hori (1964a) reported that three young out of nine ringed on Sheppey were later found dead. On the Ythan, the percentage of ducklings which fledged successfully was measured by observing the whole estuary every day at low tide, when the broods were most active. The observers searched for all known broods, noted new ones and

Table 8.1. *Fledging success (percentage of ducklings hatched which reached 40 days of age) and number of young fledged per territorial pair on the Ythan. From Patterson et al. (in press b)*

Year	Territorial pairs	Ducklings hatched	Percentage fledging	Fledged per pair
1962	71	173	19	0.45
1963	70	156	41	0.91
1964	72	126	12	0.21
1970	62	182	37	1.10
1971	76	168	9	0.20
1972	72	136	19	0.36
1973	61	221	66	2.39
1974	56	219	51	2.00
1975	60	158	47	1.23
1976	63	212	57	1.92
1977	82	334	36	1.48
1978	70	255	17	0.63
1979	65	173	19	0.49
Mean	68	193	34.6	0.99 ± 0.2

recorded the number of ducklings in each brood. Disappearance of a duckling was taken to mean mortality. Since brood-mixing affected many of the broods it was not possible to estimate the daily mortality rate of individual pairs' own ducklings, only the rate for the ducklings currently in their care and the rate for all the broods in each nursery area. The data for the whole season were used to estimate overall fledging success, the percentage of ducklings hatched which reached 40 days of age.

Fledging success varied considerably from year to year, from nine to 66 per cent with a mean of 34.6 per cent (table 8.1). The fledging success of ducklings thus seems to be much more variable than the hatching success of eggs and clutches and this poses questions as to the cause of the variability.

Most of the ducklings which disappear do so while quite young. The number of surviving Ythan ducklings decreased most rapidly in the first week of life and survival after three weeks of age was high (figure 8.1). This pattern of increasingly good survival with age is fairly general among Anatidae.

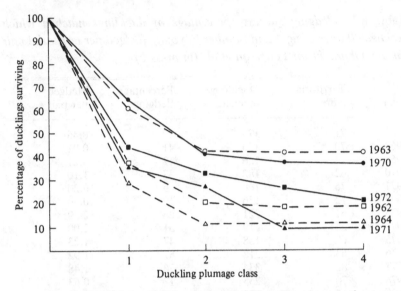

Figure 8.1 Survivorship curves for ducklings. The plumage classes are defined in table 8.14 and the numbers of ducklings hatched are shown in table 8.1. Drawn from data in Young (1964*a*) and Williams (1973).

8.2 Predation

It is rarely possible to observe the deaths of ducklings directly; most merely disappear from one day to the next. Hori (1964*a*) considered great black-backed gulls to be important predators on Sheppey and Jenkins *et al.* (1975) blamed gulls and stormy weather for the deaths of ducklings at Aberlady. Pienkowski & Evans (in press *b*) recorded much predation by herring gulls. Williams (1973) saw seven ducklings killed by herring gulls and two by crows. Some of the ducklings taken by gulls had previously been separated from their parents by human disturbance or by aggressive interaction between pairs. Young (1964*a*) also considered that scattering of ducklings by the attacks of adult shelducks made them vulnerable to predation. Mammal predators may also take some ducklings. One ringed Ythan female (and one duckling) disappeared from a brood in 1977 and the female's body was later found outside a known fox den. Hori (1969) found several dead ducklings apparently killed by foxes, stoats and weasels *Mustela nivalis* in areas with thick cover near the water's edge.

The rates of observed predation are very low. In 554 brood-hours of observation (plus some hundreds of hours spent on brood censuses) on the Ythan in 1975–7, only two cases of predation were seen. In both, a

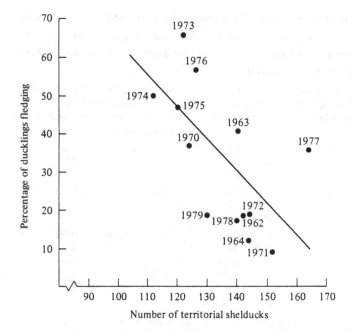

Figure 8.2 Fledging success of ducklings, in relation to the total number of territorial shelducks on the Ythan (estuary plus freshwater pools). The year is shown against each point and the numbers of ducklings hatched are shown in table 8.1. The line is the calculated regression ($y = 140.82 - 0.80x$, $r = 0.623$, $p < 0.05$). From Patterson *et al.* (in press *b*).

herring gull swooped to grab one of a scattered group of ducklings. In addition, two dead ducklings were found during the three years, one in a field 100 m from the estuary (Makepeace & Patterson, 1980).

8.3 Duckling survival in relation to density

The percentage of ducklings on the Ythan which survived from hatching to fledging decreased significantly with increasing density of territorial pairs, including those on freshwater pools (figure 8.2). The relationship was less significant when only estuarine territories were considered, and fledging success was not correlated with the size of the non-territorial flock. This suggests that the number of broods arriving on the estuary might affect survival, since many of the pairs with freshwater territories brought their young to the Ythan. It also suggests that duckling mortality might be related on a day-to-day basis with the density of broods in the nursery area.

Table 8.2. *Correlations and partial correlations between the percentage of ducklings which disappeared from the Ythan during each 24 hour period and the number of broods present at the start of the period. From Makepeace & Patterson (1980)*

Year	Number of broods Mean	Maximum	Kendal correlation coefficient		Partial coefficient controlling for weather	
1975	7.5	11	τ	−0.09	Temperature	−0.03
			DF	28		27
			P	0.26		—
1976	12.7	20	τ	+0.38	Min. temp.	+0.45
			DF	44		43
			P	0.001		—
1977	14.1	20	τ	−0.13	Temperature	−0.002
			DF	38		37
			P	0.13		—
1978	8.5	12	τ	+0.53	Wind	+0.55
			DF	32		31
			P	0.001		—

Densities of broods per hectare of the Ythan were calculated for each day for each nursery area (figure 7.7) separately, and for the whole estuary, from the number of broods present and the actual areas used by the broods that year. On the estuary, as a whole, the percentage of ducklings under ten days of age which disappeared in each 24 hour period increased significantly with increasing number of broods present at the start of the period in 1976 and 1978 (table 8.2). In both years there was a sudden increase in the daily mortality rate, from usually under four per cent to over ten per cent, at the highest densities in that year (figure 8.3). The increase occurred above 18 broods in 1976 and, more strikingly, above seven broods in 1978. In 1975 and 1977 there was no such relationship (table 8.2) and daily mortality rates varied erratically, with no tendency to increase at high density.

The difference between years remained when correlations between daily duckling mortality and weather factors (see section 8.5) were allowed for by partial correlation; indeed the coefficients were increased in 1976 and 1978 and reduced in 1975 and 1977, so accentuating the differences (table 8.2). Even in the year with the best correlation (1978) the coefficient was not large; a Pearson correlation suggested that under 40 per cent of the variation in duckling mortality could be explained by brood density ($r = +0.62$; $r^2 = 0.38$; DF = 32).

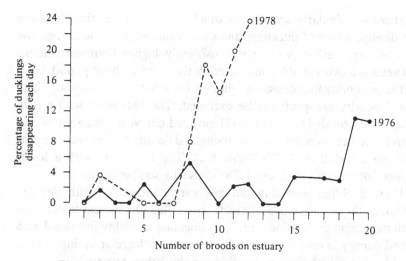

Figure 8.3 The percentage of ducklings under 10 days of age which disappeared in each 24 hour period, in relation to the total number of broods on the Ythan estuary at the start of the period. Re-drawn from Makepeace & Patterson (1980).

The disappearance of ducklings might be related to the total number of ducklings present on the estuary rather than to the number of broods. However, neither measure of density had a consistently higher correlation with disappearance of ducklings. Indeed the number of ducklings was so highly correlated with the number of broods (Kendal $\tau = 0.63$–0.89; $P < 0.001$ in all years), that these two factors could not easily be separated.

The lack of correlation between duckling density and mortality, in two of the four years, might have resulted from taking the estuary as a whole and ignoring local variations in density. Consequently the main nursery areas, Sleek, Snub, Logie and Lagoons (figure 7.6) were analysed separately (where there were sufficient data) in 1975, 1976 and 1977. Within any one area the number of ducklings present at any one time was small, so that mortality had to be calculated for ducklings of all ages, not just for those under ten days of age.

In none of the resulting nine sets of data was there a significant correlation between duckling mortality and density (largest Kendal $\tau = +0.20$; DF $= 30$; $P = 0.11$).

A number of other analyses, e.g. considering only the early part of the season up to the peak in the number of ducklings, and re-analysis excluding a few extreme points, also failed to show any correlation between duckling disappearance and density in 1975 and 1977.

There was similarly no evidence of a lag effect. Correlations between the disappearance of ducklings and density on the previous day, or one or two days earlier, were not consistently higher than correlations between the two variables measured in the same 24 hour period.

The possibility that density might only be important in years with high brood density can probably be excluded. The two years with a high number of broods (1976 and 1977) included one year with a significant correlation between duckling mortality and density (1976) and one year with no correlation (1977) (table 8.2). The two years with a lower number of broods (1975 and 1978) showed a similar variation.

Thus, duckling mortality did not vary consistently with density. Although there was an overall tendency for fledging success to decrease with increasing population size, daily duckling mortality increased with brood density in only two out of four years and the relationship did not hold for individual nursery areas within the Ythan estuary.

8.4 Duckling survival in relation to aggressive interaction

Ducklings are often attacked by adults and may be scattered far from their parents during aggressive interactions between brood pairs, as I have already described. It seems possible that such treatment might affect the ducklings' chances of survival and this possibility prompted an investigation of the relationship between duckling mortality and aggressive interaction on the Ythan (Williams, 1974; Makepeace & Patterson, 1980).

Williams (1974) found that ducklings which were involved in brood-mixing survived significantly less well than those which remained in discrete broods during their development (figure 8.4). Two different methods of estimation of mortality were used for ducklings in mixed broods but both showed a significantly lower survival for mixed ducklings, particularly in their first 15 days. Williams' study did not show what aspect of brood-mixing might be responsible for the increased mortality, but adult aggression seemed a likely possibility. This was tested on the Ythan in 1975 and 1976 (Makepeace & Patterson, 1980).

Nursery areas containing a number of broods were watched by two observers for three-hour periods and all the aggressive interactions between the brood pairs were counted. The percentage of ducklings which disappeared was measured for 24 hour periods which contained a three-hour observation of the frequency of aggressive interaction between broods. In 1975, disappearance of ducklings increased significant-

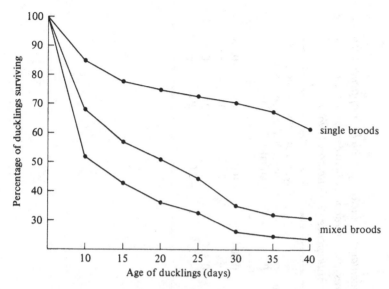

Figure 8.4 Survivorship curves for ducklings reared in single and mixed broods. From Williams (1974).

ly with increasing frequency of interaction, though not with increasing frequency of physical attacks on ducklings (table 8.3). In 1976 there was no correlation between mortality and frequency of interaction and in neither year were the results greatly affected by partial correlation controlling for any effect of density.

An alternative analysis of the same data, based on the individual broods, was possible in 1976 but not in 1975 when too few separate broods were observed. Since the interactions involving a particular brood were not independent of those of the neighbouring ones, a different measure of interaction, the number of physical attacks suffered by the ducklings of each brood, was used. The percentage of ducklings in a brood which disappeared during their first ten days of life increased significantly with increasing frequency of physical attacks suffered by the ducklings (table 8.4). This result differs from that of the previous analysis for 1976 (based on all interactions and disappearances in whole areas) but the difference is difficult to interpret since the two analyses differ in the way the data are used. The difference is unlikely to be caused by density effects since again the correlations were only slightly reduced by partial correlation analysis (table 8.4).

The relationship between duckling mortality and the amount of aggressive interaction between broods was thus somewhat inconsistent,

Table 8.3. Correlations and partial correlations between the percentage of ducklings which disappeared from a nursery area of the Ythan in 24 hours and the frequency of aggressive interactions between broods, in the same area in a three-hour observation during the same period. From Makepeace & Patterson (1980)

	All interactions per brood per hour			Physical attacks on ducklings per duckling per hour		
	Kendal τ	DF	P	Kendal τ	DF	P
1975						
Kendal correlation coefficient	+0.51	16	0.004	+0.02	16	0.462
Partial coefficient[a]	+0.38	15	—	−0.15	15	—
1976						
Kendal correlation coefficient	−0.05	20	0.396	−0.01	20	0.485
Partial coefficient[a]	−0.13	19	—	−0.04	19	—

[a] Controlling for density (broods per hectare).

Table 8.4. *Correlations and partial correlations between the percentage of ducklings which disappeared from a brood in their first 10 days and the frequency of attacks suffered by the ducklings of that brood in at least one three-hour observation during those 10 days. Data from 1976 only. From Makepeace & Patterson (1980)*

	Kendal correlation coefficient			Partial coefficient controlling for density		
	τ	DF	P	τ	DF	P
All attacks on ducklings per duckling per hour	+0.47	18	0.004	+0.37	17	—
Physical contacts with ducklings per duckling per hour	+0.46	18	0.006	+0.33	17	—

varying between years and between methods of analysis. A particular problem was that, in 1976, measurement based on mortality within areas showed no relationship between mortality and interaction while measurement based on losses from individual broods showed a significant one. One of these two conflicting results, the one based on broods is probably the less reliable since mortality was measured over a ten-day period which may have included only one (though usually more) three-hour observation of the frequency of aggression. Thus, due to inconsistencies between years, neither brood density nor aggressive interaction can completely explain variations in the disappearance of ducklings. The importance of other factors must be considered and, of these, weather was thought to be the most likely to affect duckling mortality.

8.5 Duckling survival and weather

Early mortality of ducklings increases in bad weather in a number of duck species (Koskimies & Lahti, 1964; Hilden, 1965; Bengtson, 1972; McAloney, 1973). Makepeace (1973) found in the shelduck that the daily mortality rate of ducklings increased significantly with increasing values of a chill factor, calculated from rainfall, windspeed and minimum air temperature.

Adverse weather, if sufficiently extreme, might kill ducklings directly by chilling (Koskimies & Lahti, 1964) but may more often have an indirect action by increasing the rate of energy loss. The ducklings' yolk reserves last only for a few days (Kear, 1965), during which time the young birds must learn to feed efficiently enough to start gaining energy. Any factor which increases their rate of heat loss will increase the rate at which their food reserves are used up and may in addition decrease feeding time through an increased need for brooding. Low temperature and rain may also decrease the availability of invertebrates on the mud surface (Goss-Custard, 1969). When the ducklings' net reserves (energy intake minus losses) become very low, they may die, either through direct starvation or, more probably, by becoming more vulnerable to predation. If chill factors were to increase duckling mortality in this way, through an effect on energy balance, a delay would be expected between a worsening of weather and an increase in mortality. A one-day lag effect has been found in shelducks (Makepeace, 1973) and in eiders *Somateria mollissima* (Mendenhall, 1975).

The possible relationship between duckling survival and weather has been investigated on the Ythan (Makepeace & Patterson, 1980). Ideally, weather data should have been collected on the estuary to reflect the precise conditions under which the ducklings were living. However, this was not practicable and data were taken mainly from a small meteorological station at the Aberdeen University Field Station at Culterty, 200 m from the lower Ythan estuary. Any data not obtainable at Culterty were taken from the Meteorological Office station at Dyce, Aberdeen, 17 km from the estuary. Although the microclimate on the mudflats undoubtedly differed to some extent from that at the meteorological stations, Mendenhall (1975) showed that there was a close correlation between data from the different sites.

The weather factors chosen were those likely to affect the rate of heat loss by chilling: temperature, rain and windspeed. Since the mortality rate among the ducklings was measured most commonly over 24 hour periods starting at 0800–1000 h, the weather data were taken from corresponding 0900–0900 h periods as follows:

temperature – to the nearest 0.1 °C, at Culterty at 0900 h;
minimum temperature – also to the nearest 0.1 °C, at Culterty over the 24 h and collected at 0900 h on the next day;
rainfall – to the nearest 0.1 mm, at Culterty over the 24 h and collected at 0900 h on the next day;
windspeed – to the nearest 0.1 m s^{-1}, at Dyce, using the mean of hourly readings for the 0900–0900 h period.

Since chilling would tend to increase with higher rainfall and wind speed but decrease with higher temperatures, the original temperature data were inverted to measures of 'temperature deficit' by subtracting from 20 °C.

An index of total chill effect was obtained by combining the separate weather factors. Ideally such a measure should incorporate the precise way in which heat loss increases with each of the weather factors (Leutz & Hart, 1960). However, the effects of weather factors on the rate of heat loss must be determined empirically for each species and this has not been done for shelduck. As a crude first step, a chill index was calculated by adding together the values for three weather factors so that their numerical ranges and maxima were of a similar order, by the following formula:

index $= T + R + (2 \times W)$
where T = temperature deficit (20 − temperature °C)
       R = rainfall (mm)
       W = wind speed (m s^{-1}).

In addition to these measures of daily variation in weather, an overall mean of the daily values was calculated for the main duckling period (15 May–14 July) for years when there had been an estimate of overall survival of ducklings from hatching to fledging. For each year the number of days in the duckling season when each weather factor had a higher value than a 17-year mean (from the mid-May to mid-July period, 1960–76) was calculated, since a given season's mean could be made up of varying proportions of high and low extremes of weather.

The weather data were analysed on the Aberdeen University Honeywell 66 computer, by correlation between weather factors and the percentage disappearance of ducklings in the same 24 hour period or, when examining possible lag effects, in subsequent periods up to three days later. Non-parametric correlation methods (Siegel, 1956) were used since some factors (e.g. number of broods or date) occurred as ranks of discrete values and so were not continuously distributed, and because some of the measurements may not have had a normal distribution. The Kendal correlation coefficient test was preferred since it allows the use of partial correlation analysis to control for the effect of a third factor on the correlation between two variables, although a statistical significance cannot be assigned to the partial coefficient (Siegel, 1956). Such partial correlation was necessary because many of the variables were inter-correlated. In particular, the weather factors

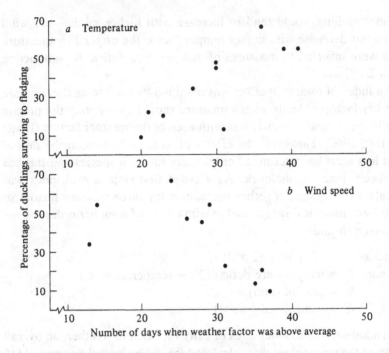

Figure 8.5 Overall duckling survival on the Ythan in relation to the number of days in the season (15 May–14 July) when the day-mean of a weather variable was greater than the 17-year mean from 1960 to 1976. Survival data are from 1962 to 1964 and 1970 to 1976. From Makepeace & Patterson (1980).

often correlated with each other and decreased together with date. At the same time duckling density usually strongly increased with date, making it difficult to separate the effects of different factors on duckling mortality.

There was no significant correlation between overall duckling survival from hatching to fledging in a given year and the mean value of any weather factor for that season (highest Kendal τ (temperature) = 0.34, $P = 0.088$). However, duckling survival increased significantly with increasing number of warm days in the season, i.e. number of days when the daily mean temperature was above the overall mean for the 17 years from 1960 to 1976 (figure 8.5 and table 8.5). Similarly, overall duckling survival decreased significantly with increasing number of windy days in the season. There was no significant correlation between duckling survival and the number of rainy days in the season or the number of ducklings hatched that year (table 8.5). The correlations were not

Table 8.5. *Correlations and partial correlations between overall duckling survival from hatching to fledging in a given year and the number of days in the season when each weather variable was above a 17-year mean value. From Makepeace & Patterson (1980)*

Variable	Kendal τ^a	P	Partial correlation controlling for				
Temperature	+0.61	0.007	wind	+0.51,	rain	+0.66
Wind	−0.58	0.010	temp.	−0.47,	rain	−0.56
Rain	+0.30	0.131	temp.	+0.43,	wind	+0.22
Number of ducklings hatched	+0.32	0.104				

[a] 8 degrees of freedom in all cases.

substantially altered by partial correlation analysis although the correlation between duckling survival and temperature was slightly increased when controlling for rain.

There is thus some evidence that overall duckling survival was affected by overall weather in the season, which raised the possibility that daily mortality might be related to the immediate weather.

There was considerable variation between years in the extent to which daily duckling mortality was correlated with weather, in which factors were involved and in the degree of lag between weather changes and subsequent changes in mortality (table 8.6). In 1975, the daily disappearance of ducklings increased with lower temperature at 0900 h on the same day and with increasing wind speed two days before. In 1976, none of the weather factors was significantly correlated with duckling mortality except if brood density was controlled by partial correlation, when mortality increased with lower minimum temperature one day before. Duckling mortality in 1977 increased significantly with lower temperature at 0900 h on the same day, with lower minimum temperature two days before and with both higher wind speed and higher values of the combined weather index three days before. However, the correlation between mortality and the weather index was greatly reduced by partial correlation controlling for brood density (table 8.6). In 1978, duckling mortality increased significantly with increasing wind speed on the same day, and with lower temperature at 0900 h, higher rainfall and higher values of the weather index one day before. Altogether, nine significant correlations were found among the 16 possible combinations of four weather factors in four years.

Table 8.6.*Correlations and partial correlations between the percentage of ducklings which disappeared in each 24 hour period and weather factors in the same period or in earlier periods (lag). Non-significant correlations have been omitted. From Makepeace & Patterson (1980)*

Year and factor	Lag (days)	Kendal correlation coefficient			Partial correlation controlling for density (broods per hectare)	
		τ	DF	P	τ	DF
1975 temperature	0	+0.26	45	0.029	+0.24	44
1975 wind speed	2	+0.23	45	0.048	+0.23	44
1976 min. temperature	1	+0.14	54	0.107	+0.30	53
1977 temperature	0	+0.24	58	0.019	+0.20	57
1977 min. temperature	2	+0.32	58	0.002	+0.24	57
1977 wind speed	3	+0.26	58	0.011	+0.20	57
1977 weather index	3	+0.22	58	0.028	+0.10	57
1978 wind speed	0	+0.21	57	0.043	+0.27	56
1978 temperature	1	+0.36	57	0.002	+0.28	56
1978 rain	1	+0.34	57	0.005	+0.27	56
1978 weather index	1	+0.41	57	0.001	+0.35	56

Thus, in each of the four years, variations in the daily disappearance of ducklings were significantly correlated with daily variation in one or more of the weather variables, although different ones were involved in different years. Low temperature, either at 0900 h or the 24 hour minimum, was involved each year and wind in three of the four years, but rain was important only in 1978. Most of the correlation coefficients were only slightly reduced when the effect of brood density was allowed for by partial correlation analysis and two (between mortality and minimum temperature in 1976 and wind speed in 1978) were increased.

There was evidence of a lag effect, in that several of the correlation coefficients reached their maximum values when mortality was related to weather on the previous day rather than to weather in the 24 hour period when the ducklings had actually disappeared. More of the coefficients reached a maximum with a one-day lag, compared to longer lags of two or three days.

A one-day lag effect of weather on duckling mortality is consistent with the hypothesis of indirect action of weather chill on mortality by increasing the rate of energy loss and thereby increasing the chance of mortality some time later. The possibility that bad weather might kill ducklings directly and quickly, say by 'exposure', is largely excluded. It

Table 8.7. *Correlations between weather factors and duckling mortality* [c] *in relation to mean weather for the duckling period in each year. Mean values greater than the four-year mean are in bold. Temperatures are deficits (from 20 °C). From Makepeace & Patterson (1980)*

	Weather factor	Mean over four years	1975	Mean in each year 1976	1977	1978
a	Temperature (°C)	7.2	6.9[a]	6.6	**7.6**[a]	**7.7**[a]
	Min. temp. (°C)	11.9	**13.4**[b]	11.1[c]	**12.1**[a]	11.6
	Rain (mm)	1.5	0.8	0.8	1.3	**3.2**[a]
	Wind (m s^{-1})	4.3	4.3[a]	3.6	**4.8**[a]	**4.7**[a]

b	Weather factor mean value in relation to four-year mean	Correlation between weather and mortality	
		significant	non-significant
	above mean	6	1[b]
	below mean	2	6

Fisher exact test $p = 0.032$

[a] Statistically significant.
[b] $p = 0.053$.
[c] Significant only when controlling for density.

is difficult, however, to explain why different factors had lags of different length and especially why the same factor differed in its lag between years. One possibility is that the degree of lag may depend on the severity of the factor in any given year, since this is known to vary (table 8.7). The difference in the lag periods of different factors explains why the combined weather index was rarely correlated with duckling mortality at any given lag. Indeed, allowing for density effects, the weather index was significantly correlated with duckling mortality only in 1978, which was the only year when even two of the weather factors (temperature and rain) reached their maximum correlation coefficients with the same degree of lag.

This study showed clearly that the daily rate of disappearance of ducklings tended to increase significantly in poorer weather. The inconsistency between years is perhaps not surprising, since the mean values of the weather factors varied considerably between years. For example, the mean daily rainfall in the duckling period was four times higher in 1978 than in 1975 or 1976 (table 8.7), and a given weather factor may not have any effect on mortality until it exceeds a threshold value. This was tested by comparing the distribution of significant correlations between mortality and weather with the distribution of mean values for the weather variables in the different years. There was a highly significant tendency for duckling mortality to increase with poorer weather only in years when that weather factor was worse than the mean for the four years' duckling periods (table 8.7b). There were no exceptions to this in 1977 and 1978, and the exceptions in the other years were either close to significance (min. temperature, 1975), only significant when controlling density (min. temperature, 1976) or occurred when the year mean values were close or equal to the overall mean (temperature and wind, 1975). Thus it appears that duckling mortality was not correlated with a given weather factor in years when the factor was 'better' than average, which might indeed suggest a threshold below which the factor had no effect.

8.6 The relative importance of different mortality factors

Since the mortality of young shelducks increased to some extent with brood density and aggressive interaction and also with adverse weather, we might ask which of these factors has the most important effect and to what extent they jointly explain variations in mortality.

Duckling mortality on the Ythan increased with increasing density in only two years out of four and with increasing aggressive interaction in only one year of two; whereas mortality increased with adverse weather factors in all four years studied. The significant Kendal correlation coefficients (τ), controlling for other factors, were +0.45 and +0.55 between mortality and density, +0.38 between mortality and frequency of aggression and ranged from +0.20 to +0.35 between mortality and weather factors, with similar sample sizes. Thus although density and aggressive interaction did not have a consistent effect on duckling mortality, in years when there were significant correlations, these tended to be higher than those between mortality and weather.

The extent to which variations in both density and weather could together explain variations in duckling mortality can be estimated from multiple regression analysis, though the results must be viewed cautiously since parametric statistical methods may not be valid with these data. The proportion of the variation in duckling mortality (r^2) which was explained by a combination of density and three weather variables was 15, 33, 6 and 47 per cent in the four years 1975–8 (table 8.8). This analysis confirmed the earlier ones in showing that density was important only in 1976 and 1978, when it explained 15 and 38 per cent of the variation in mortality, compared to 2 and 0.1 per cent in 1975 and 1977. Conversely, the weather factors were more important in 1975

Table 8.8. *Multiple regression of the percentage of ducklings disappearing in 24 hours on brood density and weather variables. From Makepeace & Patterson (1980)*

Year	Factor	Multiple regression coefficient (r)	r^2	Increase in r^2
1975	rain	0.30	0.09	0.09
	wind	0.35	0.12	0.03
	temperature	0.36	0.13	0.01
	density	0.39	0.15	0.02
1976	density	0.38	0.15	0.15
	min. temp.	0.57	0.32	0.17
	rain	0.57	0.33	0.01
	wind	0.57	0.33	0.003
1977	temperature	0.21	0.04	0.04
	wind	0.23	0.05	0.01
	rain	0.24	0.06	0.01
	density	0.24	0.06	0.001
1978	density	0.62	0.38	0.38
	wind	0.69	0.47	0.09
	rain	0.69	0.47	0.003
	temperature	0.69	0.47	0.0002

and 1977, although in these years less of the variation in duckling mortality was explained by all the factors together (table 8.8). Again, this analysis confirmed that weather factors were generally less well correlated with mortality than was density (when significant), and then explained only six to 18 per cent of the total variation in mortality.

It is clear that in all four years, and especially in 1975 and 1977, there were other important factors affecting duckling mortality. There are no data on other mortality factors, but there are a number of possibilities. Mendenhall (1975) suggested that the frequency of predation may vary with changes in the number of predators and variation in their alternative food supplies. It is also possible that duckling food supply varies in ways not related to brood density or weather. Although the identity of other mortality factors is not clear, we can probably eliminate those which varied seasonally. When the effects of brood density and weather factors were controlled by partial correlation any apparent correlation between mortality and date in the season disappeared completely ($\tau = 0.002$ to -0.06).

Since simple effects of weather on duckling mortality are not likely to change with brood density, and since density effects on mortality did not occur every year, it does not seem likely that density-dependent mortality is consistently important in limiting the production of fledglings by breeding pairs in the Ythan population of shelduck.

8.7 Strategies for rearing ducklings: single or mixed broods?

Since a large proportion, around two-thirds, of the shelducklings hatched fail to fledge, different rearing strategies adopted by the parents could greatly affect the number of independent offspring they produce. In particular, mixing of broods is common and seems to be associated with increased duckling mortality. What are the overall costs and benefits of brood-mixing, involving as it does the rearing of other birds' young?

Parental care is normally shown towards an animal's own offspring or, occasionally, towards those of close relatives. The costs of care are thus fairly obviously offset by benefits in terms of genetic fitness. However, in some species care is shown towards young which are not closely related.

In waterfowl, mixing or amalgamation of broods is common (Williams, 1974). In the eider, for example, each female brings its own young from the nest to the shore but then frequently joins with other females and their young to form large creches (Gorman & Milne, 1972; Mendenhall, 1975). Most females stay with the young for only a few days before abandoning them to the care of other, newly arrived, females. Red-breasted merganser *Mergus serrator* females interact aggressively with each other while caring for broods. The more dominant of two females may drive the more subordinate from her own young and eventually care for the whole mixed brood (Bergman, 1956).

In these, the shelduck and other similar cases, there is an interesting problem in that care is being shown to other parents' offspring which are not obvious close relatives, apart from their belonging to the same local population. The behaviour involved is complex, since it involves an interaction between donor and recipient individuals and may sometimes involve mutual exchange of young. It is not always clear, especially in shelduck, whether mixing is due mainly to the behaviour of donors or due mainly to recipients; i.e. are some parents shedding young to others or are some parents stealing young?

The possible benefits and costs of brood-mixing must be considered separately for donors and recipients and could affect fitness in two main ways, through effects on (a) the survival of each parent's own offspring and (b) the survival of the parents themselves.

Benefits and costs to offspring

A benefit of brood amalgamation to the survival of eider ducklings has been suggested by Mendenhall (1975), who showed that the risk of an individual duckling being eaten by a predator was much greater in small broods than in larger ones. Williams (1974) suggested that young shelducks in larger broods developed faster than those in small ones, but this data depend crucially on only two large broods in a sample of eight and one 'exceptional' case was excluded from analysis (Williams, 1973). These benefits in both eider and shelduck are to the recipient parents, although the survival of the donated ducklings should also be increased.

There may be a cost of brood-mixing to the survival of ducklings if there is a substantial reduction in the amount of parental care per duckling in larger broods. In bad weather, for example, the female may be able to brood only a limited number of ducklings (although Koski-mies & Lahti (1964) suggested that the increased insulation of the ducklings of sea-ducks, and their consequent independence of parental brooding, has allowed creching in these species). Williams (1974) showed that mixed broods of young shelduck had a substantial and highly significant reduction in survival compared to broods which remained separate, as I showed earlier.

Benefits and costs to parents

Gorman & Milne (1972) suggested that the benefit of brood-mixing to the donor female eider is to allow her to leave the young to feed and recoup her fat reserves which are substantially reduced during incubation. In many areas, such as the Ythan estuary, Aberdeenshire, the adult and duckling feeding areas are spatially separate so that the female might starve if she stayed continually with the young.

Hori (1964a,c) suggested that donor parent shelducks benefit by being free to migrate earlier, and presumably more successfully, than those which stay to rear their young. He considered that 'females which have spent a month or so incubating are scarcely in condition to make a migration of some 400 miles'. He suggested that pairs, especially ones breeding later, force their broods to leave them, by making attacks on the ducklings as I described in section 7.6. However, most of the evidence on brood-attacks suggest that these are attempts to drive off

Table 8.9. *Percentage annual survival of Ythan shelducks in relation to breeding history. Figures in brackets are sample sizes. From Patterson et al. (in press a)*

| | Percentage annual survival in shelducks which | | |
	fledged broods	hatched broods	did not hatch broods
Males	85.0 (40)	94.2 (52)	91.5 (94)
Females	89.1 (46)	87.5 (40)	87.6 (97)
Both sexes	87.2 (86)	91.3 (92)	89.5 (191)

No significant differences (χ^2 test).
Data from 1975–8, taking annual survival from 1 Jan to 31 Dec of the year in which breeding occurred.

other pairs' young and there is no evidence that pairs which lose their ducklings gain any benefit during migration. Williams (1974) showed that brood females had only a slightly (and non-significantly) lower survival than those which lost their broods. New data (table 8.9) showed no difference in survival between shelducks which reared broods, those which lost their ducklings and other breeding birds which failed to hatch ducklings. Thus, there is no evidence of any cost of rearing ducklings to the survival of parent shelducks (unless such costs are exactly offset by higher 'quality' of the brood parents). It is difficult to see a benefit to the survival of recipient parents, in either eider or shelduck.

Overall costs and benefits

To sum up, there are clear net benefits of brood-amalgamation in eiders. The donor adult benefits by being free to leave to feed, her young are protected in larger groups and the recipient benefits by lowering the risk of predation. There is thus mutual benefit and of course each female tends to be first a recipient and then a donor. It seems likely that in this species the benefits outweigh the costs.

The occurrence of brood-mixing in shelducks poses much more of a problem in that the lower survival of mixed broods was not offset by any corresponding benefit, e.g. in adult survival, except for the suggestion of faster growth rate in larger broods. The behaviour thus seems to be of considerable disadvantage to both recipients and donors (whose young will suffer mortality in mixed broods) and this problem prompted a study on the Ythan to measure more precisely the benefits and costs of brood-mixing in shelduck (Patterson *et al.*, in press *a*).

Table 8.10. *Percentage of Ythan males of differing dominance which were involved in brood-mixing as either donors or recipients. From Patterson et al. (in press a)*

	Upper half of rank	Lower half of rank	Unranked
Percentage involved in brood-mixing	45.5	31.6	40.9
n	21	19	22

No significant differences (χ^2 tests). Data from 1970, 1971, 1978 & 1979.

Brood-mixing among dominant and subordinate shelduck

The principal approach of the Ythan study was to compare the behaviour of dominant and subordinate birds, since dominants should be freer to behave in the way which benefits them most, possibly at the expense of the subordinates. The specific questions were (a) whether dominants are more often donors or recipients of young, and (b) whether donors or recipients are more successful in the number of their own offspring reared to fledging.

The relative dominance of males was determined in the winter flock, as I described in section 3.6. Two ranked males could be compared by their relative rank order but when at least one was unranked the measure of dominance they had in common, percentage of fights won, was used. Since not all of the breeding males came to the feeding stations, dominance data could be obtained on only 60.2 per cent of them (based on 133 males in 1978 and 1979). Brood-mixing was identified by checking the broods on the estuary every day. All could be identified by the colour-marks of the parents, by the facial plumage of the female or by position on the estuary and age of the ducklings. Mixing of ducklings was very rarely seen but was usually deduced from changes in the relative numbers and ages of ducklings in neighbouring broods from day to day (see also Williams, 1974). This shows only the final result of ducklings of the same age or any ducklings which die after mixing and before the next observation. However, the sample of brood-mixing which could be detected allowed comparison of the relative dominance of the parents involved.

There was no significant difference in their involvement in brood-mixing between males of different status (table 8.10). Between one third and one half of all pairs were affected by mixing, either as donors or

Table 8.11. *Relative rank of donor and recipient males in transfers of ducklings. From Patterson et al. (in press a)*

		Donor			Recipient		
		Male	Rank	% fights won	Male	Rank	% fights won
a	Dominant donates	012	5	59.7	399	—	12.5
	to subordinate	408	5	66.3	630	—	37.5
		613	14	41.9	630	—	37.5
		022	1	100.0	582	—	58.3
		052	—	83.3	023	25	9.4
		537	—	60.0	553	—	50.0
		907	4	82.4	1743	—	0.0
		712	9	74.1	894	—	50.0
		894	—	50.0	1743	—	0.0
		202	—	37.5	1743	—	0.0
		1713	—	100.0	563	40	0.0
		853	—	58.8	563	40	0.0
b	Subordinate	624	20	9.1	630	—	37.5
	donates to dominant	880	23	45.0	712	5	87.0

12 cases: 2 cases, $p < 0.05$ (Sign test).
Data from 1970, 1971, 1978 & 1979. Data on duckling exchange in 1971 & 1972 from Williams (1974).

recipients, with the most dominant males only slightly more involved than those of lower status.

Data on the relative dominance of donors and recipients were difficult to obtain since only about half of the territorial pairs hatched broods (Patterson *et al.*, 1974 and unpublished data) and only about one third of the brood-mixings detected were between males of known relative dominance (because only 60 per cent of the breeding males were of known status (see above)). In addition, three cases of mutual exchange of ducklings within the same season, and one case where the two males changed their relative dominance between seasons, were excluded. Of the remaining 14 cases of straightforward one-way transfer of ducklings between males of known relative dominance, 12 cases involved the more dominant male losing ducklings to the more subordinate (table 8.11). This result is unlikely if transfer were equally probable in either direction. There may be some non-independence in these data since some males were involved in more than one transfer, albeit with different 'opponents', but nevertheless, in a substantial majority of cases the donor was the relatively more dominant bird.

It is perhaps surprising that dominant males were more often donors of young, since I expected initially that dominants might usually drive subordinates away from their own young and so become recipients, as has been described in red-breasted mergansers (Bergman, 1956). However, the dominant male may attack more than the subordinate during an interaction and attacking parents moved further from their own ducklings than did attacked pairs (Makepeace & Patterson, 1980). A greater distance between parents and young might increase the chance of losing some ducklings to another pair, so that the parents which attack might be more likely to become donors of young.

Alternatively, broods may mix peacefully during feeding, when the males' relative dominance might not affect the outcome. Williams (1974) found that nine out of 14 cases of brood-mixing which he observed occurred without aggressive interaction.

Survival of ducklings to fledging

It was not possible to measure separately the survival of the original and adopted ducklings in mixed broods, since the marking of small young caused unacceptable mortality (Williams, 1974). It was also not meaningful to express survival as a percentage fledged of those hatched, when some of those fledged were derived from other broods. The measure used here is simply the actual number of ducklings fledged by each pair. To evaluate such data, however, it is necessary to estimate (a) differences in the initial brood sizes of donors and recipients and (b) the proportion of mixed broods made up of original and adopted ducklings.

Pairs which would eventually be recipients had somewhat larger broods (8.20 ± 1.20, $n = 5$) when they were first seen than had eventual donor pairs (6.23 ± 0.72, $n = 13$). The difference, however, was not significant and it is possible that some recipient pairs had already gained ducklings before being seen for the first time, since only one observation was made on each brood each day.

The proportion of adopted ducklings in a recipient's brood could be estimated from the number transferred into the brood in relation to the number present on the previous day (or the day before the first transfer if there were several). An average of 38.7 per cent of the ducklings in mixed broods were adopted, although this varied between years (table 8.12). This percentage, measured at the time of transfer, can only be used as an estimate of the proportion at fledging if we assume there was no differential mortality of the two categories of duckling after transfer.

Table 8.12. *Percentage of adopted ducklings in mixed broods immediately after mixing. From Patterson et al. (in press a)*

Year	Number of broods	Number of ducklings original	adopted	Percentage adopted
1975	7	51	42	45.2
1976	12	97	33	25.4
1977	16	121	92	43.2
1978	7	62	30	32.6
1979	4	38	36	48.7
Totals and mean	46	369	233	38.7

Table 8.13. *Mean number of ducklings reared to fledging by donors and recipients. From Patterson et al. (in press a)*

	Mean number of ducklings reared by adults which were			
	donors	recipients	both	neither
Mean	0.31^b	3.50	0.0	1.58
Standard error	0.31	1.38	—	0.45
n	13	6	5	38
Adjusted mean[a]	1.71	2.10		

[a] Assuming 38.7 per cent of the recipients' young were the offspring of donors.
[b] Mann-Whitney $U = 15$, $p < 0.05$ (one-tailed, since recipients were expected to have larger broods).
Data from all birds of known dominance (ranked or unranked) in 1970, 1971, 1978 & 1979.

Without marking the small ducklings it was not possible to check this assumption.

In absolute terms, recipient pairs reared ten times as many ducklings as did donor pairs (table 8.13). However, an estimated 38.7 per cent of the ducklings reared by recipients were in fact the offspring of donor pairs. If the data are adjusted for this (table 8.13) recipients still seemed to rear slightly more of their own offspring than did donors, although the difference was not significant.

The finding that the dominant donors eventually seemed to rear fewer of their own offspring than did recipients fails to support the original hypothesis that the behaviour shown by the dominants should be the most beneficial to the rearing of young. On Darwinian grounds, it seems generally unlikely that birds able to behave in an advantageous way would not do so but it is possible that dominance in winter does not confer any advantage in encounters between brood parents, or that the dominance order when fighting over young differs from that found in fighting over food. Too few incidents of brood-mixing were actually seen to allow this to be tested. I showed in section 6.6, however, that the dominants had strikingly higher hatching success than subordinates, suggesting that dominance can confer some advantage in breeding.

An alternative hypothesis is that brood-defence behaviour in the shelduck is not well adapted to high brood density. Williams (1974) suggested that in his and Hori's (1964a, 1969) studies, brood-mixing generally occurred when several broods were crowded into a restricted area. In contrast, pairs which occupied isolated positions on tributaries were able to exclude other broods and retain an exclusive feeding area (Williams, 1974). Such pairs reared more young (Makepeace & Patterson, 1980; Pienkowski & Evans in press b) and Jenkins et al. (1975) have shown that small, low-density shelduck populations breed more successfully than large, high-density ones. It thus seems possible that the defence of an area around the young is effectively territorial, adapted to a wide spacing of largely isolated broods and beneficial to duckling survival in that situation but not in crowded estuary populations.

This hypothesis raises the problem of why shelducks should crowd together in estuaries and suggests the possibility of another benefit, perhaps to adult survival, which would counteract the cost to breeding output. There may be two alternative strategies open to a young adult shelduck settling to breed; it could settle at high density in an estuary, perhaps attracted by an abundant food supply, where it might achieve good survival but poorer breeding success. A limited series of feeding experiments (see section 9.6) suggested that increased food in an estuary was followed by increased settlement of yearling shelduck. Alternatively, a young adult could settle in an isolated or low density area where breeding success might be higher. If, however, adult survival were poorer, the same overall fitness, in terms of offspring produced in a lifetime, might be similar in the two strategies leading to both being retained in the population. Further observations of breeding success, adult survival and the frequency of brood-mixing especially in low density populations are needed to test this speculative hypothesis, which will be discussed in more detail in the next chapter. However, Pienkowsi & Evans (in press b) found adult survival to be similar among dispersed and crowded breeding birds.

Figure 8.6 Duckling developmental stages. Re-drawn from an original by John Love in Williams (1974).

8.8 Development, fledging and dispersal

The ducklings which survive the various mortality agents and the risks of brood-mixing proceed through a series of developmental

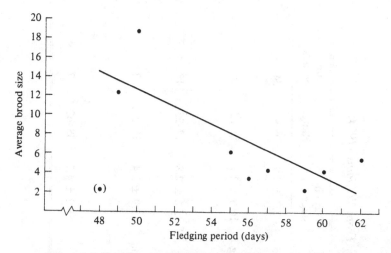

Figure 8.7 Fledging period in relation to the average size of the brood over the period. The line is the calculated regression ($r = -0.80$, $p < 0.01$) when the bracketed point is excluded. From Williams (1973). (The excluded brood contained only one duckling for most of the fledging period.)

stages until fully feathered. Williams (1974) devised a series of recognisable plumage classes (figure 8.6 and table 8.14), derived from more general ones from Southwick (1953), Gallop & Marshall (1954) and Mosby (1963). The change from class 1 to class 2 is gradual and depends mainly on an alteration in body profile, but the transition from class 2 to class 3 is marked by the fairly sudden appearance of the flank feathers. Similarly, the loss of down from the face, marking entry to class 4, occurs in a few days.

Some of the variability in the time spent in each plumage class (table 8.14) is probably due to the imprecise divisions between the classes, but Williams (1974) considered that different broods developed at different rates. He found small but non-significant differences between broods in different nursery areas which he thought might depend on differences in the abundance of food. There is also a suggestion that large broods develop faster than small ones. Williams (1974) found that larger broods spent significantly more of their time feeding than did small ones and that the total time taken from hatching to fledging was shorter for large broods (figure 8.7). Pienkowski & Evans also found faster development in a large creche. If this effect is real, it may indicate that a good food supply affects both survival and growth rate of ducklings, leading to larger broods and faster growing young. Alternatively, the best parents

Table 8.14. *Duckling plumage stages and the time spent in each. From Williams (1974)*

Class	Mosby (1963) equivalent	Description	Days in stage	Range	Number of broods
1	1a + 1b	Downy, newly hatched, patterns bright and distinct. Neck and tail not obvious. Body rounded.	7.3 ± 0.9	6–9	34
2	1c	Down colour fading and patterns becoming progressively less distinct. No contour feathers. Neck and tail become obvious. Body long and oval.	16.6 ± 2.7	13–22	26
3	11a + 11b	First feathers appear on flank and later on shoulders. Other contour feathers and secondaries develop. Face remains downy.	15.6 ± 2.8	12–23	13
4	11c + 111c	Face loses down cover. Remaining down on nape and rump gradually disappears. Predominantly feathered but cannot fly.	17.9 ± 2.4	15–23	7
5		Young able to fly.			
1–4			55.2 ± 4.7	48–62	7

may be able to protect their brood better and also ensure better feeding conditions.

On the Ythan, Young (1964a) found the mean fledging period, from hatching until the ducklings could fly, to be 58.3 ± 2–3 days in 1962–4 and Williams (1974) found a mean of 55.2 days in 1970–2 (table 8.14). Pienkowski & Evans (in press b) found that the young of isolated pairs fledged at 47 ± 1.4 days, and those of crowded pairs at 60 ± 1.6 days. Ladhams (1971) estimated a mean of 65 days for two inland broods. Boase (1951) however recorded young flying after 42–44 days, although some took over 60 days and Hori (1964a) found young flying after about 45 days. Witherby *et al.* (1939) give about 56 days as a general figure. There are obviously considerable variations between areas, which again may depend on feeding conditions.

The parent shelducks usually stay with the brood until the young can fly strongly, although Boase (1965) saw one brood of six which fledged successfully in spite of being left unattended from about one week of age in mid-July. On the Ythan, the male usually seems to leave before the female, and I have often seen broods in the care of the female alone for a week or so around fledging. Boase (1965), however, recorded several cases of males alone in charge of broods. Some cases of single parent families will be due to the death or injury of one of the partners. In one brood on the Ythan, mentioned earlier, where the female was killed by a fox, the male attended the ducklings for a few days before they joined another brood. The duration of stay by the parents may also depend on the fledging period of the young. Ladhams (1971) found that in a brood which fledged in 55 days, the parents stayed for 12 days after fledging, whereas in another, which took 75 days to fledge, the parents left 21 days before the young could fly.

After fledging and the departure of their parents on moult migration, young shelducks form flocks and move around in their local area. Gradually they disperse; many have left their natal area by mid-August (Boase, 1951; Hori, 1964a) and few remain by mid-September. Their place may be taken by juveniles from elsewhere; on the Ythan in October small groups of young, distinguishable as being from elsewhere by their lack of rings, occurred in most years.

At first the young birds do not go far. Recoveries of shelducks ringed as ducklings in Britain showed that, during July and August, 92–93 per cent of those found dead were still within their local area (figure 8.8), although two birds had travelled 45 and 50 km by August. In September, 36 per cent of the young had travelled up to 100 km, seven per cent

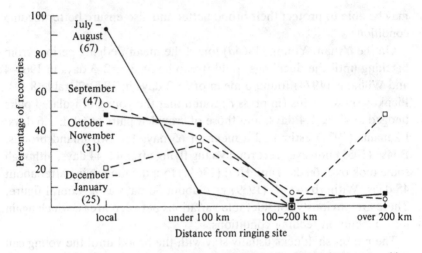

Figure 8.8 Distances moved by ringed juvenile shelducks recovered in their first autumn and winter. Sample sizes are in brackets. From data supplied by the British Trust for Ornithology.

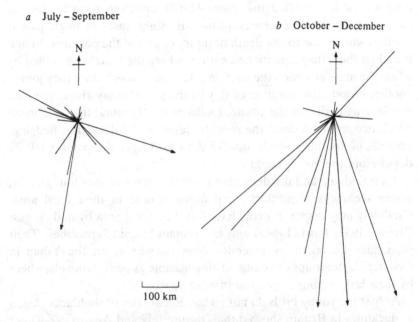

Figure 8.9 Directions and distances of recoveries of ringed juvenile shelducks in their first autumn and winter. Lines ending in arrowheads show longer-distance movement. From data supplied by the British Trust for Ornithology.

100–200 km and two birds had been recovered in western France. Over the later autumn and winter a gradually increasing proportion of juveniles were recovered at greater distances from their natal area as they dispersed for their first winter (figure 8.8). Two birds were recovered from the south of France.

Recoveries of juvenile shelducks in early autumn generally occurred in a wide spread of directions from their birthplace from southeast through south to northwest (figure 8.9a) with only a few (short distance) movements to the north and northeast. In later autumn and early winter there was a greater bias towards the south and southwest (figure 8.9b). This may have been due partly to the distribution of ringing sites, which were predominantly on the east coast of Britain, giving little opportunity for an eastward or northeastward dispersal without crossing the North Sea. However, many birds from the south coast of England, which could have gone north or east, also went southwards. It seems likely that juvenile shelducks may move south or southwest to spend their first winter in mild conditions, before returning to recruit into breeding populations, as I shall discuss in the next chapter.

9

Recruitment and the non-territorial flock

On reaching adulthood the young animal faces the problem of settling into the patch of habitat which may be its home for the rest of its life. The selection of an area is crucial to the individual's genetic fitness since it may affect both survival and breeding success and thus the total number of progeny produced in the animal's lifetime. Many individuals settle preferentially in their natal area, but others emigrate and face a choice between different regions for settlement. Selection of an area may depend mainly on assessment of the resources available but should also take into account the degree of competition for them. The response of individuals to increasing competition may limit the total number of young which settle in a local population and this may be important in the limitation of density.

In many territorial species, recruitment occurs in two stages, firstly to the non-territorial flock and then from there to the attainment of a breeding territory, sometimes several years later (Charles, 1972). In this chapter I will describe the return of young adult shelducks to the flock, the factors affecting settlement and the possible consequences for population density.

9.1 Return of the young

After dispersing for their first winter, many young shelducks return to their natal area, mostly when about one year old. Of 60 Ythan-reared ducklings (ringed before fledging) which returned, 45 per cent were first seen as yearlings, 40 per cent as two-year-olds and 15 per cent at three years old (Patterson *et al.*, in press *b*). This almost certainly underestimates the proportion returning in their first year since yearlings generally returned late in the season for only a short time, stayed in fairly inaccessible areas of the estuary and most had engraved numbered rings which were difficult to identify. All these factors would act to decrease the likelihood of yearlings being recorded in the first year that they returned. Jenkins *et al.* (1975) also found that eight Aberlady ducklings returned 'when about one year old or in their second winter . . .'.

The mean date of arrival on the Ythan of locally reared yearling females was 21 May, with little variation (range 11 April–14 June) (figure 3.2). Only one male was first seen as a yearling, on 19 May. Older birds arrived earlier in the season and I have already shown that this trend continued till at least four years of age (figure 3.2). An inability of yearlings to cope with northern winter conditions might explain why they do not return in January or February, but cannot account for them being so late in arriving. In addition, some yearlings were seen on the Ythan every winter, although these did not include any known to have been reared locally. Jenkins *et al.* (1975) similarly saw yearlings wintering at Aberlady and these also did not include any local young. It is likely that the yearlings move slowly north over the winter and spring and stay for some time in estuaries on the way. Some may not reach their natal area before the moult migration period and will not return there until two years of age.

Some of the young birds may not return at all but may settle in other areas. Such emigration is difficult to detect and measure since even ringed emigrants will be diluted in a huge unmarked population which is not observed regularly. No Ythan-reared duckling was ever seen or recovered dead in another breeding area in the breeding season (Patterson *et al.*, in press *b*). This, however, is weak evidence since few other areas were inspected and much mortality occurs outside the breeding season (see section 10.2). A better estimate of emigration was made by considering recoveries made away from the Ythan several years after ringing, to find how many of these birds had not been seen in their natal area in the interval. Only nine Ythan-reared ducklings were recovered away from the area after at least one year. Of these, three had not been recorded back on the Ythan in the $2\frac{1}{2}$, $6\frac{1}{2}$ and seven years respectively which had elapsed (Patterson *et al.*, in press *b*). It is very unlikely that they had been present but unobserved over several years since only about six per cent of adults known to be alive were missed each year (see section 10.2). It is probable that they had been resident elsewhere and thus that a proportion (one third of this small sample) of ducklings emigrate.

When the returning young shelducks arrive they join the flock of birds which have not established territories (section 4.2). At first the flock contains some late pairs, yet to become territorial (Williams, 1973) but by mid-May all the territories have been established and the young remain in a residual non-territorial group. Hori (1964*a*) described similar flocks which remained on the wintering areas on the Swale at Sheppey when the territorial pairs dispersed over the breeding areas, and Jenkins *et al.* (1975) had a persistent non-territorial, non-breeding component in the Aberlady population. However, Evans & Pienkowski (1982) found no persistent non-territorial flock at Aberlady in 1976–9, and considered that these birds had moved to large mudflats on the upper Forth.

9.2 Flock dispersion

The flock is restricted in its movements by the aggressiveness of the territorial birds and is generally confined to areas not used for territories. On the Ythan such areas are on the wider mudflats and sandbars in the middle of the upper estuary, together with some small areas between territories (figure 9.1). Similarly, Jenkins *et al.* (1975) found that the non-territorial birds at Aberlady were usually seen on the sandy areas of the estuary not used by the territorial pairs.

The Ythan non-territorial shelducks did not occur in a single flock but were scattered in several groups. Williams (1973) found that individual marked birds were highly mobile and moved frequently from group to group. Of 23 birds seen three or more times during the season, 87 per cent were seen in two or more of the flock areas in figure 9.1. Eighteen birds were seen five or more times and 89 per cent of these occurred at three sites. The non-territorial flock can thus be regarded as a single social unit distributed across a number of sites used by most or all of the birds.

Within the flock, the birds usually kept close together. Estimates of the nearest-neighbour distances between males in the Ythan flock

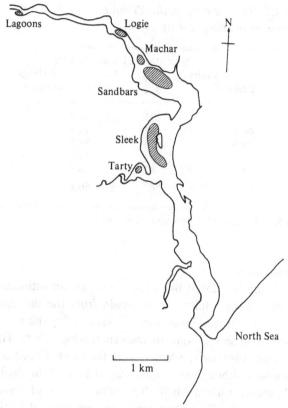

Figure 9.1 Areas of the Ythan used by the non-territorial flock.

showed that most were about 5 m apart, with few closer than 1 m and most within 10 m of the nearest other male (figure 4.3). The birds were conspicuously closer together than the territorial males at the same time.

9.3 Sex and age composition of the flock
Sex ratio

Among Ythan flock birds the sex ratio was unity, but it was variable among yearlings (table 9.1). Among 70 Ythan ducklings which returned to the area, only 26 per cent were male (Patterson *et al.*, in press *b*). Since the sex ratio among flock birds in general was equal, it seems likely that the predominance of females among the returning Ythan-born yearlings was due to greater first-year dispersal in the males, leading to fewer of them settling in breeding areas in their first year.

Table 9.1. *The sex ratio in the Ythan non-territorial flock. From Patterson et al. (in press b)*

| | Percentage of males among | | | |
| | adults | | yearlings | |
Year	Percentage	n	Percentage	n
1975	55.7	115	42.2	90
1976	49.3	152	46.8	77
1977	50.6	176	59.8	249
1978	45.5	44	19.0	42
1979	53.7	82	56.7	30
Mean	51.3	569	50.8	488

Mean for all ages, 51.1 per cent ($n = 1057$).

Age composition

As with the territorial birds (section 4.2), an estimate of age composition among the adults can be made from the distribution of minimum ages of any ringed birds present, again using the most recent year's data to minimise bias due to uneven ringing effort. The commonest ringed flock birds on the Ythan were the newly ringed adults (at least two years old) although one male was at least 19 in 1979 (figure 9.2). As with the territorial birds there was a peak of six-year-old females from the unusually high production and survival of the 1973 year class.

The generally low proportion of older birds in the flock compared to the territorial pairs (compare figure 9.2 with figure 4.1) is related to the age at which the Ythan shelducks took up territories. Of those known-age birds (ringed as ducklings or yearlings) which eventually became territorial, none did so as a yearling, half did by two years old and three-quarters by three years old (figure 9.3). About 20 per cent were not territorial in any given year even when older, but all of those which were going to take territories had done so by four years of age.

Some of the older birds which occurred in the flock were known to have been territorial and to have bred successfully in previous years. Williams (1973) recorded four such cases, including one old male which was non-territorial in three consecutive years. All four were unpaired males but later work has shown that females can also become non-territorial after having bred in earlier years.

Territorial birds may also occur occasionally with the flock (although

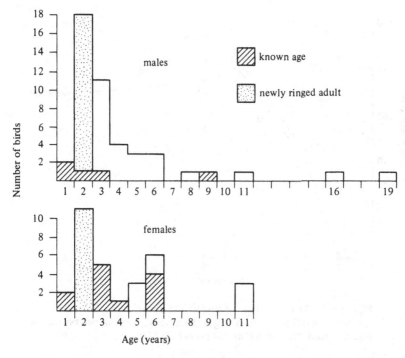

Figure 9.2 Age distribution of ringed birds in the Ythan non-territorial flock, 1979. The unshaded histograms show minimum age only (birds ringed as adults). From Patterson *et al.* (in press *a*).

they have not been included in figure 9.2). For example, in 1978, 15 of the 70 territorial pairs were seen among groups of flock birds, mostly only once but some two or three times. Williams (1973) also found that territorial and non-territorial birds associated while feeding on newly sown grain fields near the Ythan in April, although the known territorial males seemed to maintain rather larger nearest-neighbour distances than did the flock males.

Yearling shelducks, distinguishable by their plumage (section 2.2), made up almost half of the Ythan flock, with a slightly higher proportion among females than among males, both overall and in four out of the five years when data were collected (table 9.2). There was considerable variation between years, which was largely explained by variation in previous production. The proportion of yearlings in the flock was significantly higher following years when a larger number fledged (figure 9.4). Flock size, however, was only slightly (and not significantly) higher in years following higher production of fledglings (figure 9.5).

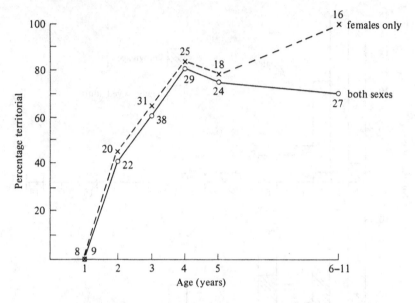

Figure 9.3 The percentage of ringed Ythan shelducks which were territorial at different known ages. Sample sizes are shown beside the points. From Patterson *et al.* (in press *b*).

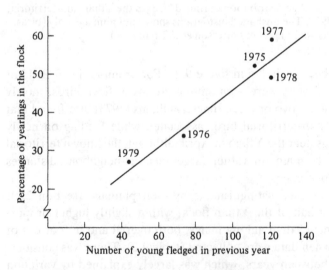

Figure 9.4 The percentage of yearlings in the Ythan non-territorial flock, in relation to the number of young fledged the previous year (shown beside the points). The line is the calculated regression ($y = 8.96 + 0.37x$, $r = 0.956$, $p < 0.02$). From Patterson *et al.* (in press *b*).

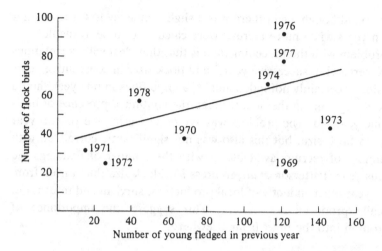

Figure 9.5 Flock size in each year in relation to the number of young fledged in the previous year (shown beside the points). The line is the calculated regression ($y = 35.0 + 0.26x, r = 0.435, p > 0.1$). From Patterson *et al.* (in press *b*).

Table 9.2. *Age composition among unmarked birds in the Ythan non-territorial flock. From Patterson et al. (in press b)*

		Percentage of yearlings among		
	males		females	
Year	Percentage	*n*	Percentage	*n*
1975	37.3	102	50.5	103
1976	32.4	111	34.7	118
1977	47.1	189	63.1	236
1978	28.6	28	58.6	58
1979	27.9	61	25.5	51
Mean	38.3	491	51.1	566

Mean for both sexes, 45.1 per cent ($n = 1057$).

There was no correlation at all between the number of fledglings produced and flock size two years later, which is consistent with most birds returning as yearlings. Variation in the survival and return of the yearlings was allowed for by calculating an index of available recruits from the total fledged multiplied by the proportion of ringed ones which survived and returned. Flock size was not significantly correlated with this index, although again there was a slight tendency towards a larger flock in years when more recruits were calculated to be available.

A problem with the preceding data is that they deal with correlations in time-series (i.e. successive years) and flock sizes in consecutive years were almost certainly not independent; a large flock in one year should lead to a large one in the next through the continued presence of birds over one year old. The problem was avoided by using the year-to-year change in flock size, but this also was not significantly correlated with the number of recruits available or with the number of fledgings the previous year (Patterson et al., in press b). Flock size thus varied from year to year independently of local production, survival and return, i.e. of locally produced recruitment. This suggests the importance of immigration into the flock from elsewhere.

9.4 Immigration into the flock

Permanent immigration into the Ythan population by shelducks reared elsewhere was difficult to detect, since ducklings were rarely ringed elsewhere and almost all immigrants would thus be unmarked. However, it was possible to test for the temporary presence of non-resident birds (transients) which might potentially become permanent immigrants (Patterson et al., in press b).

(a) It was possible to estimate the minimum number of known individuals which occurred in the study area in a given season and to compare this with the actual peak number present in May. For example, in 1979, the number of ringed birds seen or caught (205), plus the unmarked mates of some of them plus known unmarked territorial or brood pairs totalled 275 whereas the mean total count for May was only 178 and the highest single count, 205. This discrepancy suggests that there was a flow of birds through the area, unless the counts missed a substantial proportion of the population, which is unlikely given the complete coverage of the area in a short time.

(b) Some of the ringed shelducks occurred only in winter and were not seen after late March, even though they reappeared in subsequent winters. A number of these birds were seen in searches of more northern populations, suggesting that they pass through the Ythan during their return from the moult migration. Evans & Pienkowski (1982) found similar wintering birds at Aberlady.

(c) The yearlings in the non-territorial flock included a larger proportion of unmarked birds than did the previous year's locally reared

Table 9.3. *The percentage of birds with rings among fledglings and yearlings of the same Ythan year-class. From Patterson et al. (in press b)*

Year	Fledglings		Yearlings		χ^2	P
	Percentage	n	Percentage	n		
1969	50.0	122	12.8	39	15.4	<0.001
1970	83.8	68	15.4	52	52.9	<0.001
1971	60.0	15	21.4	68	4.8	<0.05
1973	63.0	146	19.6	56	28.8	<0.001
1974	28.6	112	18.8	96	0.4	>0.1
1975	33.8	74	4.1	74	19.4	<0.001
Mean	51.4	537	15.6	385	132.0	<0.001

ducklings. A proportion of each year-class of ducklings was ringed before fledging and the percentage ringed was calculated from the total known to have fledged. The percentage ringed was counted in a sample of yearlings in the flock the following year. The mean percentage ringed among the yearlings (15.6 per cent) was significantly lower than that among the previous year's ducklings (51.4 per cent) and was also significantly lower in five of the six years' data (table 9.3). This suggests the presence of a large number of unringed immigrant yearlings in the non-territorial flock, unless there was heavy differential mortality of ringed fledglings. (This is unlikely at least in years when a high proportion was ringed since the remaining unringed ones were too few to account for the number of yearlings present.)

(d) There was evidence of shelducks moving on from the Ythan. On 21 May 1969 I saw a flock of 27 ascending rapidly from the flock area and leaving to the northwest at 2015 h. This direction, if maintained, would have taken the birds to the Moray Firth which contained the next major shelduck populations 100 km to the north. Other groups were seen leaving on the evenings of 27 May and 28 May 1969 and yearlings were present in all of these flocks.

Several separate sources of information thus suggest the presence of potential immigrants at least passing through the Ythan population, although it is not possible to estimate what proportion might have settled permanently. One yearling male, which had been ringed as a duckling on the Moray Firth, settled to become territorial.

9.5 Seasonal changes in flock size

The shelducks in the Ythan non-territorial flock were counted at low tide in the afternoon when fewest were in the nesting area (section 5.1). The birds were distinguished from the territorial ones by their close spacing and their position in areas of the estuary never seen to be defended.

The number of shelducks in the flock varied considerably. In some years (e.g. 1970) the number increased steadily through May, while in others (e.g. 1978, 1979) it decreased from an early peak (figure 9.9). While some of the variation was undoubtedly due to error (e.g. birds missed or duplicated) during the counts, many of the changes in number were probably real, since birds were seen leaving (previous section). The early peaks in 1978 and 1979 occurred when most of the birds were seen feeding on newly sown grain fields near the Ythan and this additional rich food supply might have encouraged passage birds to stay. I will discuss experiments to test this in the next section.

9.6 Limitation of recruitment into the flock

In most years the size of the non-territorial flock remained stable throughout May and in two years numbers decreased, while shelducks including yearlings were seen leaving. This suggests the possibility that the number of birds in the flock might be limited, perhaps by increasing density and aggressive interaction as more birds settled.

Flock density in relation to the number present

The nearest-neighbour spacings between males in the Ythan non-territorial flock were estimated at intervals during observation periods and the birds present were counted at the same time. Three different flock areas were analysed separately since they varied considerably in size. In all three there was a decline in the mean distance between neighbouring males when there were more birds in the area (figure 9.6). As expected, the birds stayed further apart in the larger area of the Sandbars than on the small mudflats of Logie and Machar. Small nearest-neighbour spacings occurred even when few birds were present, showing a tendency for the birds to stay together, but large spacings were absent with larger numbers (figure 9.6). This suggests that as more flock birds arrive they do not spread out at the same density but are progressively crowded together, presumably because they are confined to restricted areas by the territorial pairs.

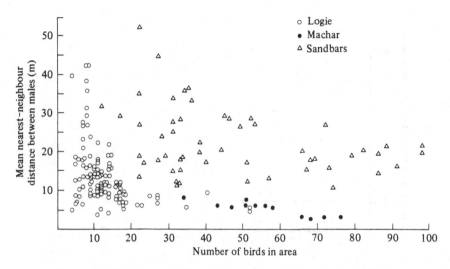

Figure 9.6 Nearest-neighbour distances between males in the Ythan non-territorial flock, in relation to flock size. The different symbols refer to separate sites of differing areas.

Aggressive interaction in relation to density

Aggressive encounters between birds in the Ythan non-territorial flock were counted, as in the winter flock, by watching groups of ten birds for five minutes. The nearest-neighbour distances in the groups were estimated at the same time and the mean distance calculated. The frequency of aggressive interaction increased significantly as the nearest-neighbour spacing decreased, i.e. as the birds became more closely crowded together (figure 9.7). There was considerable variation in the amount of aggression at any given mean spacing but the different flock areas gave similar results. The amount of interaction evidently rises as more birds arrive and crowd together. Which birds suffer most from this aggression may depend on their dominance.

Dominance

It was not possible to describe a conventional dominance hierarchy in the non-territorial flock since only a small proportion of the birds were marked. In addition, most of the late-arriving young birds were unmarked so that the ringed birds were a biased sample of the flock. It was possible, however, to distinguish yearlings from birds at least two years old, and the adult bird was the winner in 88.9 per cent of 81 aggressive encounters between males ($\chi^2 = 27.1$, $P < 0.001$). The

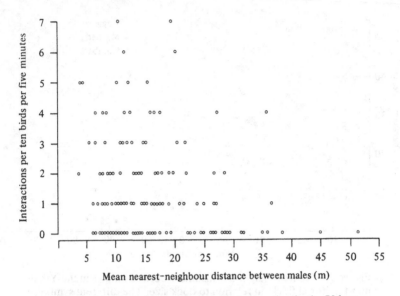

Figure 9.7 Frequency of aggressive interaction in the Ythan non-territorial flock, in relation to nearest-neighbour spacing between males. Data from Logie and Sandbars, 1976.

yearlings were thus clearly subordinate to older birds and would tend to suffer most of the consequences of aggression in the flock.

The limited-immigration hypothesis

Murton (1968) showed that subordinate woodpigeons were forced to leave flocks by the aggression of dominants as the amount of food dropped during the winter. This tended to limit flock size in relation to the amount of food while the dominant birds remained in good condition. The process was one of progressive emigration in relation to food supply. It seems possible that in the shelduck non-territorial flock, a similar effect could operate on immigration. I have shown earlier that on the Ythan there were potential immigrants, many of them yearlings, which appeared to move through the area, staying for a time in May. The non-territorial flock was confined to small areas, increased in density as more birds arrived and showed an increase in the frequency of aggressive interaction with density. Since most of this aggression was directed at the more subordinate yearlings, probably the most mobile component of the population, it seems possible that some might fail to settle as the amount of aggression reached their individual limits of tolerance of it. This process could limit flock size, in relation to

Table 9.4. *Dispersion of non-territorial shelducks over the Ythan flock areas (defined in figure 9.1) in 1977. The area with experimental feeding is in bold*

| | | | Percentage of birds in each area | | | | | |
	Tarty	Sleek	Sandbars	Machar	Logie	Lagoons	Other	n^a
a Before experimental feeding	8.2	20.7	8.2	6.1	54.5	1.7	0.6	343
b With experimental feeding	0.0	2.0	**74.2**	15.6	8.1	0.0	0.0	837

^{a} Based on repeated counts of the flock over several days.

space, food supply or other resource, by progressive limitation of immigration as competition increased. Since so few yearlings were marked, it was not possible to test whether the most subordinate were least likely to settle but it was possible to test the effects of increased food supply on flock size.

Feeding experiments
    In May of 1976 and 1977, when four or five consecutive counts of the Ythan flock had shown no further increase in its size, 7 kg of soaked wheat was scattered daily on a mudflat in one part of the flock's range and flock size and dispersion were determined regularly.

    There was no simultaneous control, which would have required a similar estuary nearby. Instead, the two experimental years were followed by two control years when no additional food was supplied. Feeding started one week later in one year than in the other, as a further guard against coincidental increase in flock size.

    The distribution of the flock changed immediately after food was supplied. In 1977 the proportion of flock birds in the section of the estuary where food was supplied changed from 8.2 per cent before feeding to 74.2 per cent afterwards (table 9.4). The number of flock birds increased abruptly and significantly within a day or two of food being supplied, from a mean of 77.6 to 108.7 (an increase of 40.1 per cent) in 1976 and from 76.8 to 121.0 (57.6 per cent) in 1977 (figure 9.8). None of the counts after food was supplied overlapped with those before and the change occurred similarly in both years in spite of the change in starting date. Evans & Pienkowski (1982), however, found no similar increase following experimental feeding of the Aberlady winter flock.

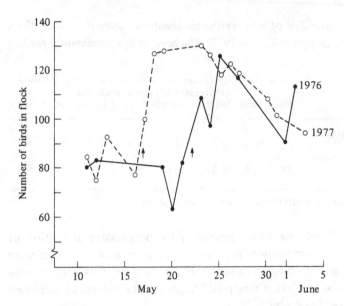

Figure 9.8 Changes in the size of the Ythan non-territorial flock with experimental supplementary feeding. The arrows show the start of feeding, which continued to the end of the period shown. From Patterson *et al.* (in press *b*).

In the control years, on the Ythan, flock size was measured before and after the dates when food was first supplied in 1976 and 1977. Using the 1976 starting date (22 May), the number of flock birds decreased in 1978 by 15.9 per cent (mean 84.5 to 71.1) and in 1979 by 1.2 per cent (68.8 to 68.0). With the 1977 starting date (16 May) there was a similar decrease of 18.9 per cent (88.3 to 71.6) in 1978 and 11.7 per cent (76.0 to 67.1) in 1979 (figure 9.9).

Thus, although there was no simultaneous control, the 1978 and 1979 results suggest strongly that the previous increases in flock size, following experimental food supply, had occurred at a time when numbers would otherwise be expected to decrease. It is likely that some of the immigrant birds, known to occur in the estuary, stayed for longer when additional food was supplied. It was not known, however, whether this would make them more likely to recruit permanently into the population by returning to stay in the following season.

9.7 Recruitment from the flock to territories

Most shelducks first became territorial at two or three years of age, i.e. one or two years after joining the flock (figure 9.3). However,

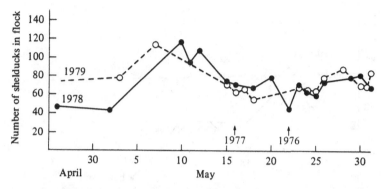

Figure 9.9 Seasonal changes in the size of the non-territorial flock on the Ythan in 1978 and 1979. The arrows show when experimental feeding was started in 1976 and 1977.

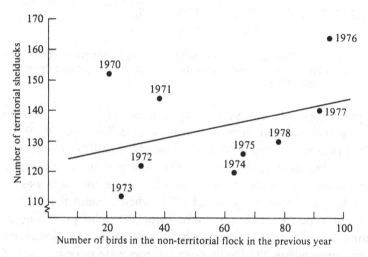

Figure 9.10 The number of shelducks with territories on or near the Ythan, in relation to flock size in the previous year (shown beside each point). The line is the calculated regression ($y = 123.36 + 0.20x$, $r = 0.332, p > 0.1$). From Patterson *et al.* (in press *b*).

the number of territorial birds in the Ythan population did not increase significantly with the number in the flock either one or two years previously (figure 9.10). Again, this analysis may not be strictly valid in time-series data, but the year-to-year change in the number of territories also showed no correlation with flock size one or two years before. The number of territories thus seemed to be determined by factors other than the number of potential recruits from the flock. This was confirmed by the long term results of the feeding experiments.

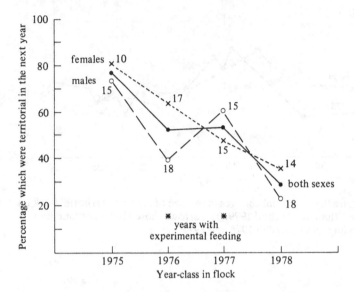

Figure 9.11 The percentage of shelduck in the Ythan non-territorial flock which became territorial in the following year. The figures show sample sizes. From Patterson *et al.* (in press *b*).

The percentage of each year's Ythan flock birds which became territorial within one year showed a progessive decline, with 70–80 per cent of the 1975 flock but only 20–35 per cent of the 1978 flock becoming territorial in the next year (figure 9.11). The percentage becoming territorial within two years showed a similar decline and there was no tendency for the flocks in 1976 and 1977, when additional food was supplied, to produce more territorial birds than the preceding or following years. This result may again have been affected by a long term fluctuation in numbers; the feeding experiments were done in years of peak numbers which were followed by a decline (see section 10.1). A decrease in the proportion of flock birds which become territorial might thus have been expected. Nevertheless, there was no indication that the increased flock sizes after experimental feeding in 1976 and 1977 had in any way altered the downward trend in the proportion of birds becoming territorial. The problem of what limits the number of territories will be discussed in section 10.6.

9.8 Recruitment strategies for young shelduck

For a yearling shelduck, the choice of a breeding area in which to settle involves a crucial decision which may greatly affect its fitness in terms of survival and breeding output. Many of course return to their natal area with which they had presumably become familiar as ducklings and newly fledged juveniles. This familiarity may confer some advantage (e.g. in knowledge of the best feeding areas) on their return, although such information could easily be gained by joining the existing residents on arrival. Return to the natal area would also be advantageous if there has been any selection on the local population to adapt individuals to the particular characteristics of the area (Wynne-Edwards, 1962). Since yearling females seem to return to the natal area more than males, the advantages of doing so may differ between the sexes. It is possible that female nesting behaviour is adapted to the local tradition of nest sites, making it more advantageous for females to breed in the area where they were reared. Some yearlings may attempt to return to their natal area but may be excluded. None of the three Ythan young which apparently emigrated was seen as a yearling but brief visits could easily have been missed.

Those birds which settle in areas other than their natal one pose a more interesting problem since they must select an area. There are two main aspects to be assessed: (a) the chance of survival, and (b) the potential production of offspring in one area as against another.

Important factors influencing survival are food supply and the competition for it. The potential recruit to an area may be able to make some estimate of food in a breeding area from the feeding rate attainable in the flock areas and from the amount of aggressive interaction there. The mobility of flock birds between several parts of an estuary may help in such assessment. A yearling shelduck moving from one breeding area to another may be able to compare them and select one to settle in the following season. A possible mechanism for selection could be an effect of food supply on length of stay (suggested by the Ythan feeding experiments) and a tendency to settle subsequently where most time was spent as a yearling. Other factors affecting survival, e.g. predators, could influence settlement in the same way, by affecting duration of stay as a yearling.

To be able to breed it is essential that the settling shelduck is eventually able to acquire a territory and the earlier this is possible the greater the possible number of years of breeding. The likely competition for territories may be assessed by the frequency of attacks on flock birds by territorial ones as the flock moves around an estuary. For example, on the Ythan in 1975, attacks by territorial on flock birds occurred at the rate of seven per hour ($n = 10$ h). Such attacks could again affect duration of stay.

The likely success of breeding may be more difficult to predict in advance. Areas with a better food supply may well give better breeding as well as better survival. Yearling females visit the nesting areas and investigate nest holes in the company of older females and so may be able to compare the nest sites in different breeding areas. Good sites may again attract longer stays.

This hypothesis on recruitment strategy for the yearling shelduck is thus that a number of factors will have positive or negative effects on the duration of an individual's stay in a particular area in the first year. (Being born in an area will of course have a major influence.) The resultant of these various effects will be variation in the length of stay in different areas, with the young bird tending to return to the place where it spent most of its time as a yearling. The major problem with this hypothesis is in testing it! It would be difficult to measure duration of stay in different areas by individual yearlings. Even within one area, the crucial transient potential settlers are unlikely to be marked before arriving and the experience of being trapped and marked might well be a major incentive to move on!

10

The limitation of
shelduck populations

As individual animals strive to increase their own genetic fitness, by attempting to increase their survival and breeding output, their behaviour will have inevitable consequences for their local population. Striving for dominance, defence of territory, interaction over nest sites or between brood parents and the decisions taken by young animals settling in an area may all affect population density and processes such as breeding and mortality rates. In this chapter I will bring together data already presented, along with material on mortality, to discuss the problem of what limits the size of local shelduck populations.

Shelduck numbers in Britain as a whole have increased during the last 20 years (section 2.7), without any sign of levelling off by 1979. The cause of this increase is not clear, although the nutrient enrichment of estuaries and more effective protection may have had an influence.

This change in the national population could have occurred by the increase of individual local populations, or by a spread of range to include areas not previously occupied, or both. There is some evidence of a spread of range (Sharrock, 1976), especially to inland sites, which can perhaps account for part of the increase. However, changes in local populations cannot be excluded and this raises the question of what

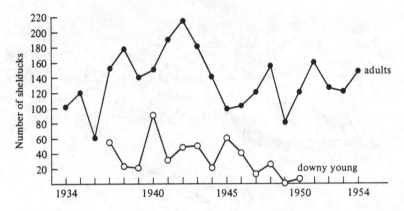

Figure 10.1 Total numbers of adult shelduck in May (maximum for
April and May, 1951–4), and downy ducklings on the Tay. From Boase
(1951, 1959).

limits local population density in shelducks. Two problems are involved:
(a) what limits the population to around a particular mean level (which
may have increased), and (b) what causes fluctuations about that level?
The same factors may or may not be involved in both.

10.1 Changes in local population size

In general, local populations of shelduck seem to have remained
fairly stable. The earliest and longest continuous series of counts were
those of Boase (1951, 1959) on the Tay estuary. The total number of
birds present in May fluctuated around 120, with no overall trend
towards increase or decrease, from 1934–54 (figure 10.1). This population seems, however, to have been lower in earlier years, having been
over 100 only once (in 1919) between 1912 and 1934 (Boase, 1951).
Similarly, long series of counts of spring numbers in several local
populations in the Firth of Forth (Jenkins, 1972; Jenkins *et al.*, 1975)
showed no general tendency toward increase in the period 1950–73
(figure 10.2). Numbers at Aberlady, however, approximately doubled
in a period without counts between 1957 and 1967. A continuing upward
trend in the number of Aberlady breeding birds from 1967–73, if
extrapolated backwards, would coincide with the counts up to 1957
(figure 10.2), which may suggest a steady increase in this population
since about 1958. Later work at Aberlady by Pienkowski & Evans (in
press *b*) shows that the breeding population remained at 130 birds
between 1977 and 1979. The mean number of shelducks present on the
Ythan in May showed a large fluctuation with peaks of about 230 in 1962

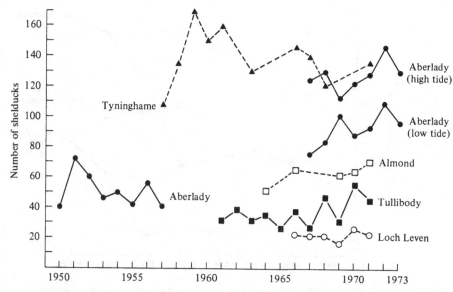

Figure 10.2 Spring counts of adult shelduck at a number of localities in the Firth of Forth and Loch Leven. From 1967, the high tide counts at Aberlady represent the total local population and the low tide counts the estuarine breeding population. From data tabulated in Jenkins (1972) and Jenkins *et al.* (1975).

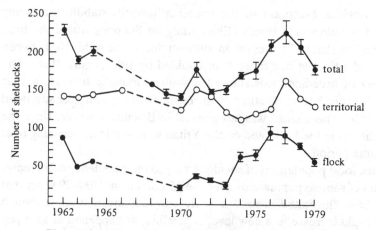

Figure 10.3 The number of territorial and non-territorial shelduck and the total population on the Ythan, 1962–79. From Patterson *et al.* (in press *b*).

and 1977 with a low of 141 in 1970 (figure 10.3). The change in number was progressive from year to year, apart from a minor increase in 1971. There was no tendency towards overall increase or decrease.

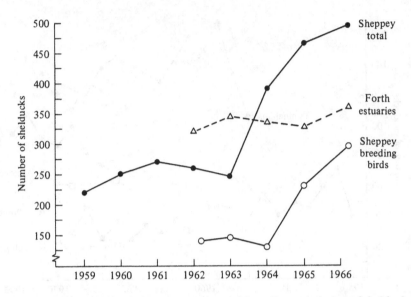

Figure 10.4 The total population and breeding population of shelducks at Sheppey, and at four localities on the Firth of Forth (Tullibody, Kennet, Almond and Tyninghame). From data in Hori (1969) and Jenkins (1972).

An obvious exception to the tendency towards stability or steady change in numbers is Hori's (1969) study on Sheppey where the total numbers of shelducks present in summer increased suddenly between 1963 and 1964 and had more than doubled by 1965 (figure 10.4). The number of breeding birds showed a similar dramatic increase a year later. It is not known whether the increase was temporary, or sustained after 1966. This change was not general in Britain, however, since the populations in the Forth and on the Ythan were stable or declining over the same period.

Thus, local populations of shelduck vary in their tendency to increase. Studies of sample populations over the 45 years from 1934–79 show that most have fluctuated around a fairly constant mean level, although fairly rapid increase to a new level is possible, as happened at Sheppey and Aberlady. In these cases, presumably, a limiting factor had been relaxed allowing increase until limitation by a new one. However, there is no general evidence of increase in local populations, supporting the idea that the national increase has probably occurred through a spread into new areas.

A comparison of the mean densities in different areas, to attempt to

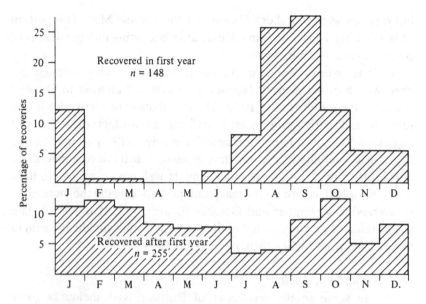

Figure 10.5 Seasonal distribution of recoveries of British-ringed shelducks found dead. Those reported as having been dead for some time have been excluded. From data supplied by the British Trust for Ornithology.

relate these to variations in resource levels, is not possible. The published data are mainly in the form of total numbers of shelducks, without the precise area in which they live being specified. In large estuaries, like the Tay, the area used may have varied from year to year, while on Sheppey the breeding birds occupied a network of freshwater creeks whose total area is difficult to compare with that of an open shore. Our interest must therefore be focused on year-to-year fluctuations in population size and the various processes which might affect them. Some of these, reproduction and the emigration and immigration of juveniles, have already been described in earlier chapters. I will discuss their relevance to change in population size after considering the remaining major process, mortality.

10.2 Shelduck mortality

Timing of mortality

Most juvenile shelducks which die do so in the first two months after fledging. Over half of the ducklings ringed in Britain and recovered in their first year were found dead in August and September (figure 10.5). There was another minor peak of recoveries in January, but very few were found dead between February and May. This pattern of heaviest recovery rate immediately after becoming independent is of course common among birds.

Adult recoveries were distributed much more evenly through the year, with a broad peak in January to March and another in October just after the moult (figure 10.5). The distribution of recoveries in the summer and early autumn may well be an artifact of the moult migration. The low number of recoveries in July and August may be due to the inaccessibility of the moulting grounds which are occupied then. Any birds dying during these months might well not be recovered until they are washed ashore on the mainland, perhaps causing the increase in recoveries in September and October (figure 10.5). If this is so, late winter remains the only period of the year when adult shelducks seem to have a somewhat higher than average mortality.

Causes of death

In some of the recoveries of British-ringed shelducks (data supplied by the British Trust for Ornithology) the finder mentioned an apparent cause of death. Many birds were reported as having been shot, which is surprising considering that the species is fully protected. Of 129 juveniles recovered in their first six months, 27.3 per cent had been shot, compared to 14.6 per cent of 398 older birds ($\chi^2 = 9.3$, $P < 0.01$). There was no difference in the proportions of juveniles reported as shot before and after 1967 but the position seems to have improved for older shelduck. Of 140 adults recovered prior to 1967, 23.1 per cent were shot, compared to only 14.4 per cent of 248 recovered in later years. This apparent difference is, however, not quite significant ($\chi^2 = 3.75$, $P > 0.05$). These percentages may not reflect the true proportion shot, since shot birds may be more likely to be recovered than those dying from other causes, while some evidence of shooting may be suppressed if the recovery is reported by the shooter.

Many other apparent causes of death were mentioned in reports of recoveries. Shelducks were found oiled (7), below cables (2), caught in barbed wire (1), killed by a dog (1), killed by predators (9, with fox, mink *Mustela vison* and peregrine *Falco peregrinus* mentioned), drowned in stake nets (1), poisoned (2), diseased (2), 'injured' (10) and, unfortunately, drowned in traps (11). (This last category will be greatly over-represented since all such birds will be found and the cause of death will be obvious. It makes up, however, a very tiny proportion of the 4,876 shelducks which had been ringed up to 1979.)

Parasites and disease have occasionally been reported in shelducks. Harrison (1957) found a female with avian tuberculosis and heavy infestations of four species of trematodes and two species of cestodes. Osieck & Roselaar (1972) reported 'numbers' of shelduck among thousands of wildfowl and waders found poisoned by botulism in the Netherlands.

Shelducks also seem to feature heavily among birds found dead in cold weather. Dobinson & Richards (1964) reported 710 shelducks found dead in the severe winter of 1962–3, a far higher total than for any other waterfowl. Similarly, Harrison & Hudson (1964) described 106 dead shelducks as being more than any other waterfowl species except wigeon *Anas penelope*. Boyd (1964) reporting 754 shelduck bodies found, suggested that 'in proportion to the numbers likely to have been present, the shelduck suffered most severely'. Pilcher (1964), counting the bodies of birds on the shores of the Wash, suggested that 'the number of shelduck found must represent a very high proportion of the total winter population . . .'. The birds found were mainly in very poor condition and very few were thought to have been shot. The shelduck's very specialised feeding method, which relies on filtering invertebrates from the surface layers of soft mud, will make it very vulnerable during severe weather when mudflats may be frozen for prolonged periods. However, Hori (1964*d*) found that eight out of nine shelducks he found dead at Sheppey in early 1963 had *Hydrobia* in their gizzards and only three of the birds were emaciated. He attributed some of the deaths to exposure and had one captive bird die in the presence of unlimited food. Young (1964*c*) found that some Ythan birds returned in 1963 without webs on their feet, suggesting frostbite and the danger of freezing to the substrate.

Estimation of survival rate

Estimates from recoveries of ringed birds. I have analysed the recoveries of shelducks ringed in Britain, using Haldane's maximum-likelihood method for incomplete data (Murton, 1966). Birds ringed with aluminium rings (prior to 1962) were excluded to minimise bias due to ring loss and birds which were long dead when found were also not used. Analysis was carried out on the Aberdeen University Honeywell computer (using programmes written by Dr S. R. Baillie) and gave an estimate of 69.3 ± 2.6 per cent mean annual survival for birds ringed as adults (table 10.1) and 62.1 ± 2.8 per cent for those ringed as ducklings (table 10.2, including recoveries in the first year). These analyses were

Table 10.1. *Interval between ringing and recovery in British-ringed shelducks, ringed when full-grown with monel rings. From data supplied by the British Trust for Ornithology*

Year ringed	Greatest possible interval (years)	Total recovered	Interval between ringing and recovery (years)														
			1	2	3	4	5	6	7	8	9	10	11	12	13	14	15
1962	18	6	4	—	—	—	—	—	—	1	—	—	—	—	—	—	1
1963	17	17	5	2	1	—	2	3	1	—	1	1	—	—	—	1	—
1964	16	3	2	1	—	—	—	—	—	—	—	—	—	—	—	—	—
1965	15	5	1	—	1	—	—	1	1	—	1	—	—	—	—	—	—
1966	14	5	2	1	1	—	—	—	—	—	—	1	—	—	—	—	—
1967	13	1	—	—	—	—	—	—	—	—	—	—	—	—	1	—	—
1968	12	6	1	3	1	—	—	1	—	—	—	—	—	—	—	—	—
1969	11	20	12	3	1	1	1	2	—	—	—	—	—	—	—	—	—
1970	10	40	25	2	2	3	2	1	3	1	1	—	—	—	—	—	—
1971	9	9	6	1	—	1	—	—	1	—	—	—	—	—	—	—	—
1972	8	7	4	1	2	—	—	—	—	—	—	—	—	—	—	—	—
1973	7	10	5	—	2	1	—	1	1	—	—	—	—	—	—	—	—
1974	6	3	1	—	—	—	2	—	—	—	—	—	—	—	—	—	—
1975	5	8	5	1	—	—	1	—	—	1	—	—	—	—	—	—	—
1976	4	2	1	—	—	1	—	—	—	—	—	—	—	—	—	—	—
1977	3	8	3	2	3	—	—	—	—	—	—	—	—	—	—	—	—
1978	2	11	8	3	—	—	—	—	—	—	—	—	—	—	—	—	—
1979	1	4	4	—	—	—	—	—	—	—	—	—	—	—	—	—	—
Total		165	89	20	14	7	8	9	7	3	3	2	0	0	1	1	1

Table 10.2. *Age at recovery of British-ringed shelducks, ringed as ducklings with monel rings. From data supplied by the British Trust for Ornithology*

Year ringed	Greatest possible age at recovery	Total recovered	Age-class at recovery (years)												
			1	2	3	4	5	6	7	8	9	10	11	12	13
1959	21	2	2	—	—	—	—	—	—	—	—	—	—	—	—
1960	20	1	—	1	—	—	—	—	—	—	—	—	—	—	—
1961	19	2	2	—	—	—	—	—	—	—	—	—	—	—	—
1962	18	10	7	—	—	—	—	—	—	—	—	—	1	1	1
1963	17	16	10	—	1	—	1	—	—	1	—	2	1	—	—
1964	16	3	3	—	—	—	—	—	—	—	—	—	—	—	—
1965	15	5	4	—	1	—	—	—	—	—	—	—	—	—	—
1966	14	7	4	—	—	—	—	1	1	—	—	1	1	—	—
1967	13	9	4	1	1	—	1	—	1	1	—	1	1	—	—
1968	12	11	9	—	1	—	—	—	1	—	—	—	—	—	—
1969	11	21	16	1	1	1	1	1	—	1	—	—	—	—	—
1970	10	15	12	—	1	—	—	1	2	—	—	1	—	—	—
1971	9	8	3	—	1	1	1	1	—	—	—	—	—	—	—
1972	8	7	5	—	—	—	1	1	2	1	—	—	—	—	—
1973	7	10	6	1	—	—	1	1	1	—	—	—	—	—	—
1974	6	7	6	—	—	—	1	—	—	—	—	—	—	—	—
1975	5	1	1	—	—	—	—	—	—	—	—	—	—	—	—
1976	4	4	4	—	—	—	—	—	—	—	—	—	—	—	—
1977	3	8	8	—	—	—	—	—	—	—	—	—	—	—	—
1978	2	5	5	—	—	—	—	—	—	—	—	—	—	—	—
1979	1	5	5	—	—	—	—	—	—	—	—	—	—	—	—
Total		157	116	4	5	7	5	5	7	3	0	5	2	1	1

based on years of survival calculated from the exact dates of ringing and recovery, and there were minor variations when survival was calculated on the basis of calendar years (or from 1 July for birds ringed as ducklings). For example, adult survival based on calendar years was estimated at 77.6 ± 2.5 per cent. It is not clear why there should be this difference, since about 60 per cent of shelducks ringed as adults (and later recovered) were caught in January to March, near the start of the calendar year (from data supplied by the British Trust for Ornithology).

These survival estimates give an expectation of life of 2.7 years for adults and 2.1 for juveniles, using Murton's (1966) method of calculation. Boyd (1962) estimated mean adult survival as 80 per cent per year, and life expectancy at 4.5 years.

Estimates from the disappearance of colour-marked birds. The survival of adult Ythan shelducks was estimated by considering the birds seen or ringed after 1 January in one year and finding the percentage of these recorded after 1 January of the next year (the return rate). Disappearance could of course be due to emigration rather than death but such movement was considered separately (see below). It was possible for a bird, actually alive, to be missed in a given year. Six per cent of 451 birds (known to be alive from subsequent sightings) were missed for at least one year and the survival estimates for these years were adjusted accordingly.

Newly ringed birds, in their first year after ringing, showed a consistently lower return rate than previously ringed ones (figure 10.6) with a greater difference between the two categories in 1974–5 to 1977–8 than in the earlier years. Return of newly caught birds was higher in years when more of them were caught late in the season; these birds had a shorter interval in which to die and were probably less likely to be transients than birds caught in winter.

Shelducks other than those newly ringed had a mean annual return rate of 77.8 per cent, varying from 67 to 86 per cent between years (figure 10.6). The year-to-year fluctuations tended to follow those among the newly ringed birds (except for 1974–5), suggesting a common factor causing real changes in survival.

Territorial Ythan shelducks had a consistently high rate of return, varying only from 85 to 93 per cent with a mean of 88.6 per cent (figure 10.7), which was slightly higher in the later years of the study. This gives a mean expectation of life of 8.3 years. Return was similar in territorial males (90.9 per cent of 186) and in females (88.0 per cent of 183,

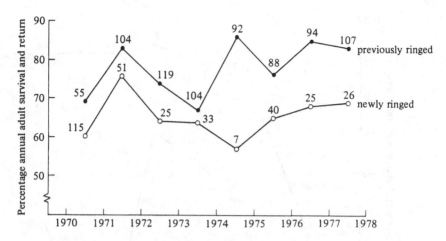

Figure 10.6 Percentage annual survival and return of newly ringed and previously ringed adult shelduck at the Ythan in different years. The first and last years of study have been excluded as less reliable. The figures are sample sizes. From Patterson *et al.* (in press *b*).

$\chi^2 = 0.02$). Jenkins *et al.* (1975) found that five out of eight shelducks breeding at Aberlady in 1971 were still alive in 1973 and 90 per cent of 21 breeding in 1972 survived to 1973.

Non-territorial shelducks had a much lower and more variable return rate than had territorial ones (figure 10.7). The difference may have been caused by greater mobility and emigration of the non-territorial birds; it seemed not to be due to newly ringed birds since the low return of non-territorial shelducks in 1975–6 and 1976–7 (figure 10.7) coincided with an increased return of newly ringed birds in those years (figure 10.6).

These survival estimates based on the disappearance of marked Ythan adults are generally higher than those made from the recoveries of ringed birds found dead, even though Ythan birds made up almost 60 per cent of the national recovered sample. Unless survival is unusually good on the Ythan, this may suggest that shelducks which are likely to be recovered after death have a lower survival than average.

Survival and return in the first year of life was estimated by the percentage of ducklings, ringed at 30–40 days old, which were recorded in the study area one or more years later. The return rate varied greatly between years, from 8–32 per cent, with a mean of 18.7 per cent (table 10.3), and was best for the 1973 year-class which also had the largest number of fledglings produced. There was, however, no overall correla-

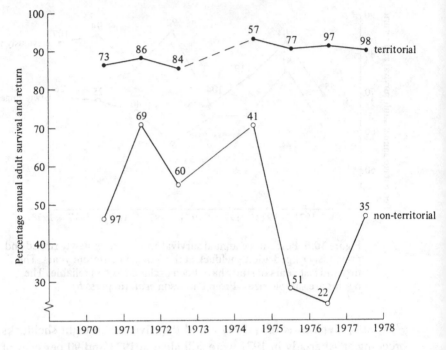

Figure 10.7 Percentage annual survival and return of territorial and non-territorial adult shelducks at the Ythan in different years. The figures are sample sizes. From Patterson *et al.* (in press *b*).

tion between return rate and the number of fledglings produced. Yearling return was not correlated with that of adults in the same year, possibly because the two categories disperse separately in winter.

Emigration. Some of the disappearance of colour-marked birds may of course have been due to their settling elsewhere. Such emigration was difficult to estimate since even ringed emigrants would be diluted in a huge unmarked population which was not being observed regularly. No territorial or breeding adult or Ythan-reared duckling was ever seen or recovered dead in another breeding area in the breeding season. However, this is weak evidence since few other areas were inspected and most mortality occurred outside the breeding season (figure 10.5). A better estimate was made from recoveries away from the Ythan at least one whole year after ringing, to find how many of these birds had been absent for at least one breeding season before

Table 10.3. *Survival and return of Ythan-ringed duck-lings in their first year of life. From Patterson et al. (in press b)*

Year of ringing	Number ringed	Number alive in next year[a]	Percentage return
1968	80	9	11.3
1969	61	5	8.2
1970	57	13	22.8
1971	9	2	22.2
1972	7	1	14.3
1973	92	29	31.5
1974	32	9	28.1
1975	25	3	12.0
1976	59	8	13.6
Total	422	79	18.7

[a] Including those which were not seen in their first year but which were seen subsequently.

being recovered. Of 23 birds ringed as adults and recovered away from the Ythan at least one year later, only eight (35 per cent) had not been seen on the Ythan in each intervening year. Only three had never been seen between ringing and recovery; the other five had been seen occasionally but always in winter. All eight were males, whereas of the 15 which had returned regularly to the Ythan, only nine were males ($P = 0.050$, Fisher test). The eight 'non-residents' were recovered in a wide variety of places; only three of the eight were recovered on the moulting grounds near northwest Germany and two were recovered on the west coast of Scotland whereas of the 15 'residents', ten were recovered on the moulting grounds and the other five on the east coast of Scotland, on the presumed return migration route.

None of the eight shelducks which were absent from the Ythan before being recovered was a territorial or breeding bird before disappearing, suggesting that there was little emigration of established residents. The data above are consistent with the occasional catching and ringing of transient birds, mainly males, on the Ythan in winter. The inclusion of such birds in the non-territorial sample may explain the lower return rate of this category. I have already discussed the emigration of Ythan juveniles (section 9.1), estimated at around one third of those fledged.

10.3 Survival and fluctuations in population size

It is possible that a slow increase in local population size, such as that at Aberlady in 1967–73, could have resulted from increased survival, especially of non-territorial adults or juveniles. However, the survival of territorial breeding adults is so high that changes in it cannot account for major changes in the size of a breeding population, such as happened at Sheppey between 1964 and 1966 (figure 10.4). At the Ythan, where survival was measured each year, variations in population size were not correlated with variations in survival.

A decrease in population size might be expected after a severe winter, since large numbers of shelducks are found dead then. However, the Tay population increased from 1946 to 1947 (Boase, 1951), in spite of the intervening hard winter. Similarly, a group of four local populations on the Firth of Forth also showed an increase between 1962 and 1963, after the severe winter (figure 10.2). Shelduck numbers on Sheppey dropped only slightly between these two years and the reduction was no greater than between 1961 and 1962 (figure 10.4). Some populations, such as the Ythan (figure 10.3) and Tyninghame and Tullibody on the Forth (figure 10.2), showed decreases between 1962 and 1963 but these were within the normal range of fluctuation seen in other years without severe winters. An index of the British wintering population also showed no large decrease after the 1962–3 winter, although the overall upward trend in population was halted for several years (figure 2.8). Thus, there is no general and convincing evidence of a decrease in shelduck populations following hard winters. This is very surprising in view of the large numbers of birds found dead in severe weather and suggests that most of the mortality may have been suffered by winter immigrants or non-breeding stocks, which would not be counted in censuses of breeding populations. The reduction in the rate of increase of the national winter index after 1962–3 perhaps suggests that the high losses reduced the establishment and growth of new populations rather than having an effect on established ones.

10.4 Breeding

The output of independent juveniles from a shelduck population could be limited by the proportion of birds which breed, by hatching success or by fledging success. To have a regulatory effect on population size, these variables should be density dependent and the output of young should have an effect on subsequent population size.

Breeding success in relation to density
Young (1964a, 1970b) suggested that a considerable proportion of adult territorial shelducks did not breed each year. However, I have argued (section 6.8) that there is little evidence for this, and it seems that most or all territorial pairs at least start to lay. Any reduction in output of young must occur during the nest and duckling stages.

On the Ythan, hatching success, fledging success and the number of young fledged per territorial pair all decreased significantly with increasing number of territorial pairs in the population (sections 6.6, 8.3 and 8.7). Hori (1969) also showed that on Sheppey the number of ducklings hatched per breeding pair was lower after a large increase in population size. Pienkowski & Evans (in press b) found at Aberlady that fledging success decreased significantly as the number of broods increased, and that isolated pairs had markedly higher breeding success than had pairs in Aberlady Bay (see section 8.7). Jenkins *et al.* (1975) found that a number of populations of shelduck produced a higher ratio of young per adult at lower population levels. There thus seems to be a fairly general tendency for a density-dependent reduction in breeding output with increasing population size in shelducks.

Breeding output and subsequent population size
In the Ythan population, there was no significant correlation between the number of fledglings produced and changes in the size of the non-territorial flock or of the total population in the following year or two years later. Population size also did not vary with changes in the calculated number of available local recruits, based on production, survival and return of yearlings (section 9.3). However, flock size and total population size did tend to increase from 1974 (figure 10.3) following high production survival and return of young from 1973. Boase (1951) interpreted a high population on the Tay in 1942 as a result of good breeding in 1940, and Hori (1969) considered that the sudden rise in the Sheppey total population in 1964 and in the breeding population in 1965 followed from a high production of ducklings in 1963. At Aberlady, a peak in the population size in 1972 followed unusually high fledgling success in 1969, and Evans & Pienkowski (1982) found that winter numbers were correlated with production of fledglings three years earlier. It may be that, although, in general, variations in breeding output are not correlated with subsequent changes in population size, exceptional output may be followed by a temporary increase in numbers one to three years later.

The lack of correlation between change in population size and either survival or breeding output suggests that emigration or immigration may be important in causing some of the variation in numbers in local populations. The presence of potential immigrants has been shown in the Ythan population (section 9.4) and some immigration may be necessary to maintain numbers where populations are not self-sustaining.

10.5 Are estuarine shelduck populations self-maintaining?

Jenkins *et al.* (1975) considered that at Aberlady '. . . not enough birds were reared locally in most years to maintain the resident population. The stability of the numbers of residents must therefore have usually been due partly to immigration of birds bred elsewhere.' Similarly, Hori (1969) recorded an increase of 155 in the Sheppey breeding population between 1963 and 1964 following the hatching of 515 ducklings in 1963. Applying the mean Ythan value of 34.6 per cent survival from hatching to fledging, about 178 young may have fledged. To replace adult mortality and also to produce the observed increase, most of these would have had to survive their first year and return to their natal area, whereas on the Ythan only 18.7 per cent of fledglings survived and returned. It thus seems likely that much of the dramatic increase on Sheppey was due to immigration.

From the data on reproduction and survival on the Ythan it is possible to test whether the population could maintain itself through local production. With a mean annual disappearance of 11.4 per cent among a mean of 136 territorial birds, there were an estimated 16 deaths each year. A mean production of 67 fledglings, with 18.7 per cent (of ringed ones) returning, gives only an estimated 13 locally reared recruits to replace the lost adults. There would then be further disappearance among the young in the non-territorial flock before they became territorial. With 24.0 per cent disappearance between the first and second years ($n = 25$) and 22.3 per cent between the second and third years ($n = 47$), the 13 returning young were probably reduced to eight new territorial adults to replace the 16 lost each year.

Since three-quarters of returning ducklings were females, six out of the eight estimated new territorial recruits each year may have been females. These, however, were still fewer than the eight female deaths (half of the 16 losses, since territorial males and females had similar survival).

It thus seems clear that returning locally produced young were too few to replace adult mortality in the Ythan population. Some additional young, however, (possibly about one-third of the survivors; section 9.1) may have settled elsewhere. These additional survivors should be added to those which returned, assuming that a similar proportion of young in other populations would be available to settle on the Ythan and that some mutual exchange of recruits occurs between different local populations. If such emigrant young are added, the estimated mean number of survivors is increased from 13 to about 20, which would produce about 12 territorial adults two years later (calculated as before). This number is still inadequate to replace the annual losses, although possible errors in the various estimates which form the calculation make it difficult to be certain that the Ythan population is not self-maintaining. In particular, the estimate of emigration is unsatisfactory. There is, however, no indication that any surplus of young is produced, except possibly in years of unusually high production (e.g. 1973). Since recruitment, on average, barely balanced losses, some immigration (probably of more males than females) must have occurred in many of the years of the study to maintain the observed stability of the population and to explain the lack of correlation between locally reared recruitment and subsequent changes in population size.

If other similarly stable populations of shelduck are not self-sustaining through local production every year, there is a problem in identifying where the necessary immigrants come from. If different populations vary in breeding output independently of one another, young may move from an area with surplus production to make up a deficit elsewhere. In a number of neighbouring estuaries in the Forth there was no tendency for the production of fledglings to vary in parallel, with peak numbers of young occurring in different years in the different places (Jenkins, 1972). None of these estuaries had unusually high production in 1963 when there was a large output of young on Sheppey (Hori, 1969). It is thus possible that immigrants to one population may originate from a surplus produced in another in the same year. There may also be shelduck populations which more often produce a surplus of young over their requirement to replace adult mortality. Jenkins *et al.* (1975) found that a number of small and low-density populations in the Firth of Forth produced more young per pair than did larger denser populations. Within Aberlady, they also found that the best production was from relatively isolated broods occupying the upper stream rather than the main estuary. Similarly, on the Ythan, broods in a narrow tributary were more successful than those elsewhere (Makepeace & Patterson,

1980). Isolated pairs on the shores of the Forth in 1972–3 produced from 2.0–6.2 young per pair, well above the average for estuaries (Jenkins *et al.*, 1975). Pienkowski & Evans (in press *b*) later confirmed this by finding that such isolated pairs produced 0.72–1.04 young per pair compared to only 0.12–0.32 young per pair for pairs in Aberlady Bay. The survival of isolated pairs' ducklings was about six times higher than that of the Bay ducklings. Adult mortality was similar in the two areas.

It thus seems possible that the fairly large numbers of shelducks breeding in small groups or at low density, or even as isolated pairs, may be producing more young than are needed to replace adult mortality. These surplus young may recruit into the estuarine populations and maintain their numbers in years when their own production has been low.

10.6 Limitation of population size

It seems likely from the foregoing discussion that immigration is the most important factor influencing year-to-year fluctuations in shelduck populations, at least the most-studied estuarine ones. Variations in mortality seem to have little effect on subsequent numbers, and changes in breeding output seem only to be important in years of exceptionally good breeding success. Local population size may thus be limited by factors limiting recruitment through immigration.

Potential recruits to a population must first settle in the non-territorial flock and then establish a territory before they can breed. Recruitment may thus be limited at two levels: entry to the flock and attainment of a territory.

Flock size may be limited by competition among the non-territorial birds for resources such as food or space. The Ythan feeding experiments (section 9.6) suggest that increased food supply can increase numbers, possibly by inducing more potential immigrants to stay. Competition may also affect mortality or the likelihood of emigration by established flock birds. The competitive situation and chance of survival elsewhere will also be important, which may explain why more recruits settle in years after good breeding, when there will be more competition everywhere and a greater 'immigrant pressure' on any given area.

The number of breeding pairs in an area will depend on the demands for space by individual territorial birds. Territories seem only to be established where there is above a minimum abundance of food. The most likely individual function of the territory in shelducks is the provision of an exclusive feeding area for the female during pre-laying,

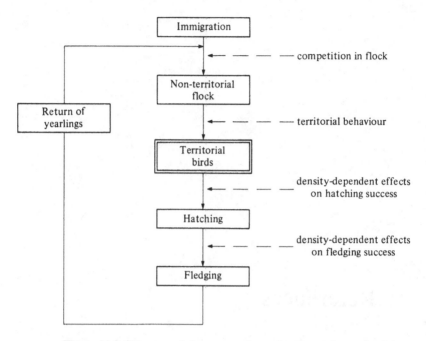

Figure 10.8 Diagram of the stages of recruitment to the territorial breeding population in shelducks (boxes and arrows) and the points at which limitation could occur (at the right).

laying and incubation. Territory size will depend on the amount of space required around the female to prevent interference by other shelducks. This area is likely to be independent of variations in food abundance above the minimum required by the pair. The number of breeding pairs in a shenduck population will depend on the total area available which has at least the minimum food density. This hypothesis suggests that increases in the breeding population, such as occurred at Sheppey between 1964 and 1965 (figure 10.4), should have involved the colonisation of new territorial areas where presumably the food supply had risen above the threshold for territory establishment. I similarly suggested earlier that the general increase in numbers of shelducks in Britain probably involved mainly the colonisation of new areas.

This hypothesis of limitation of population size at two levels, recruitment to the flock and recruitment to territories, together with density-dependent effects on hatching and fledging success (figures 6.7 and 8.2), is likely to regulate numbers and may explain the considerable stability of shelduck populations.

References

This list includes, for completeness, some papers which are not cited in the text. Some minor references to shelduck, especially records of occurrence, and references in general works and local faunas, have been omitted.

Alexander, W. B. & Lack, D. (1944). Changes in status among British breeding birds. *British Birds*, **38**, 62–9.

Allen, R. H. & Rutter, G. E. (1956). The moult-migration of the Shelduck from Cheshire in 1955. *British Birds*, **49**, 221–6.

Allen, R. H. & Rutter, G. E. (1957). The moult-migration of the Shelduck from Cheshire in 1956. *British Birds*, **50**, 344–6.

Allen, R. H. & Rutter, G. E. (1958). The moult-migration of the Shelduck from Cheshire in 1957. *British Birds*, **51**, 272–4.

Allen, R. H. & Rutter, G. E. (1963). The Shelduck population of the Mersey area in summer 1957–63. *Wildfowl Trust Annual Report*, **15**, 45–6.

Ashcroft, R. E. (1976), A function of the pairbond in the Common Eider. *Wildfowl*, **27**, 101–5.

Atkinson-Willes, G. L. (1969). The mid-winter distribution of wildfowl in Europe, northern Africa and south-west Asia, 1967 and 1968. *Wildfowl*, **20**, 98–111.

Atkinson-Willes, G. L. (1976). The numerical distribution of ducks, swans and coots as a guide in assessing the importance of wetlands. *Proceedings of the International Conference on Wetlands and Waterfowl*, Heiligenhafen, 199–271.

Atkinson-Willes, G. L. & Salmon, D. G. (1977). Wildfowl censuses and counts in Britain. *Wildfowl*, **27**, 162–3.

Atkinson-Willes, G. L. & Scott, P. (eds.) (1963). *Waterfowl in Great Britain*. HMSO, London.

Bannerman, D. A. & Lodge, G. E. (1957). *The Birds of the British Isles*. Volume 6. Oliver & Boyd, Edinburgh & London.

Bateson, J. D. (1968). *A study of territorial behaviour in relation to food supply in the Shelduck (Tadorna tadorna Linn.) population of the Ythan estuary*. Unpublished MSc dissertation, Aberdeen University.

Bauer, K. M. & Glutz von Blotzheim, U. N. (1968). *Handbuch der Vögel Mitteleuropas*. Volume 2. Akademische Verlagsgesellschaft, Frankfurt.

Baxter, E. V. & Rintoul, L. J. (1953). *The Birds of Scotland*. Oliver & Boyd, Edinburgh.

Bengtson, S-A. (1972). Reproduction and fluctuations in the size of duck populations at Lake Myvatn, Iceland. *Oikos*, **23**, 35–58.

Bergman, G. (1956). Om Kullsammanslagning has skrakar, *Mergus serrator* och *Mergus merganser*. *Fauna och Flora*, **1956**, 97–110.

Bertram, B. C. R. (1978). Living in groups; predators and prey. In *Behavioural Ecology*, eds. J. R. Krebs & N. B. Davies. Blackwell, Oxford.

Blake, E. (1975). First recorded occurrence of the European Shelduck in the southern hemisphere. *Ostrich*, **46**, 258.

Boase, H. (1935). On the display, nesting and habits of the Shelduck. *British Birds*, **28**, 218–24.

Boase, H. (1938). Further notes on the habits of Sheld-duck. *British Birds*, **31**, 367–71.

Boase, H. (1950). Notes on the behaviour of some ducks. *Scottish Naturalist*, **62**, 1–16.

Boase, H. (1951). Sheld-duck on the Tay estuary. *British Birds*, **44**, 73–83.

Boase, H. (1959). Shelduck counts in winter in East Scotland. *British Birds*, **52**, 90–6.

Boase, H. (1965). Shelduck broods on the Tay estuary. *British Birds*, **58**, 175–9.

Boyd, H. (1953). On encounters between wild White-fronted Geese in winter flocks. *Behaviour*, **5**, 85–129.

Boyd, H. (1962). Population dynamics and the exploitation of ducks and geese. In *The Exploitation of Natural Populations*, eds. E. D. Le Cren & W. M. Holdgate. Blackwell, Oxford.

Boyd, H. (1964). Wildfowl and other water-birds found dead in England and Wales in January–March 1963. *Wildfowl Trust Annual Report*, **15**, 20–2.

Breckenridge, W. J. (1956). Nesting study of Wood Ducks. *Journal of Wildlife Management*, **20**, 16–21.

Brown, J. L. (1963). Aggressiveness, dominance and social organisation in the Steller Jay. *Condor*, **65**, 460–84.

Bryant, D. M. (1978). Moulting Shelducks on the Forth Estuary. *Bird Study*, **25**, 103–8.

258   *References*

Bryant, D. M. (1981). Moulting Shelducks on the Wash. *Bird Study*, **28**, 157–8.
Bryant, D. M. & Leng, J. (1975). Feeding distribution and behaviour of Shelduck in relation to food supply. *Wildfowl*, **26**, 20–30.
Bryant, D. M. & Waugh, D. R. (1976). Flightless Shelducks on the Forth. *Scottish Birds*, **9**, 124–5.
Burt, W. H. (1940). *Territorial behaviour and populations of some small mammals in southern Michigan.* University of Michigan Zoological Miscellaneous Publications, **45**.
Buxton, N. E. (1975). *The feeding behaviour and food supply of the Common Shelduck (Tadorna tadorna) on the Ythan Estuary, Aberdeenshire.* Unpublished Ph.D. Thesis, Aberdeen University.
Buxton, N. E. (1976). The feeding behaviour and food of the Shelduck on the Ythan Estuary, Aberdeenshire. *Wildfowl*, **27**, 160.
Buxton, N. E. (1981). The importance of food in the determination of the winter flock sites of the Shelduck *Tadorna tadorna*. *Wildfowl*, **32**, 79–87.
Buxton, N. E. (in press). Territorial use and feeding behaviour in the breeding of the Common Shelduck *Tadorna tadorna* L. *Ornithologische Gesellschaft*.
Buxton, N. E. & Young, C. M. (1981). The food of the Shelduck in north-east Scotland. *Bird Study*, **28**, 41–8.
Calhoun, J. B. (1962). Population density and social pathology. *Scientific American*, **206**, 139–48.
Campbell, J. W. (1947). The food of some British Wildfowl. *Ibis*, **89**, 429–32.
Charles, J. K. (1972). *Territorial behaviour and the limitation of population size in Crows, Corvus corone and C. cornix.* Unpublished PhD Thesis, Aberdeen University.
Clapham, C. S. & Stjernstedt, R. (1960). Ringing young Shelducks. *The Ringers Bulletin*, **1**, 3–4.
Cody, M. L. & Cody, C. B. J. (1972). Territory size, clutch size and food in populations of Wrens. *Condor*, **74**, 473–7.
Coombes, R. A. H. (1949). Sheld-duck: migration in summer. *Nature*, **164**, 1122–3.
Coombes, R. A. H. (1950). The moult migration of the Shelduck. *Ibis*, **92**, 405–18.
Coulson, J. C. (1966). The influence of the pair bond and age on the breeding biology of the kittiwake gull *Rissa tridactyla*. *Journal of Animal Ecology*, **35**, 269–79.
Coulson, J. C. (1968). Differences in the quality of birds nesting in the centre and at the edge of a colony. *Nature*, **217**, 478–9.
Cramp, S. & Simmons, K. E. L. (eds.) (1977). *The Birds of the Western Palearctic*. Volume I. Oxford University Press, Oxford.
Davies, N. B. (1978). Ecological questions about territorial behaviour. In *Behavioural Ecology*, eds. J. R. Krebs & N. B. Davies. Blackwell, Oxford.
Davies, N. B. (1980). The economics of territorial behaviour in birds. *Ardea*, **68**, 62–74.
Davis, J. (1975). Breeding behaviour of two species of duck. *Nature in Wales*, **14**, 206.

Davis, T. A. W. (1975). Behaviour of Shelduck with young. *Nature in Wales*, **14**, 206–7.

DeFries, J. C. & McClearn, G. E. (1970). Social dominance and Darwinian fitness in the laboratory mouse. *American Naturalist*, **104**, 408–11.

Dementiev, G. P. & Gladkov, N. A. (eds.) (1952). *Birds of the Soviet Union*. Volume 4. Moscow.

Dewhurst, F. W. (1930). Field notes on the Sheld-duck. *British Birds*, **24**, 66–9.

Dobinson, H. M. & Richards, A. J. (1964). The effects of the severe winter of 1962/63 on birds in Britain. *British Birds*, **54**, 373–434.

Drent, R. & Swierstra, P. (1977). Goose flocks and food finding: field experiments with Barnacle Geese in winter. *Wildfowl*, **28**, 15–20.

Dunnet, G. M. (1966). Social organisation in shelduck, *Tadorna tadorna*, on the Ythan estuary. *Proceedings of the Royal Society Population Study Group*, **2**, 45–8.

Dzubin, A. (1955). Some evidences of home range in waterfowl. *Transactions of the North American Wildlife Conference*, **20**, 278–98.

Eltringham, S. K. (1961). The moulting Shelduck of Bridgwater Bay. *Somerset Birds Report*, **47**, 59–61.

Eltringham, S. K. & Boyd, H. (1960). The Shelduck population in Bridgwater Bay moulting area. *Wildfowl Trust Annual Report*, **11**, 107–17.

Eltringham S. K. & Boyd, H. (1963). The moult migration of the Shelduck to Bridgwater Bay, Somerset. *British Birds*, **56**, 433–44.

Enkelaar, H. & Lebret, T. (1966). The seasonal distribution of hatching dates of the Shelduck in the S.W. of the province of Zeeland, The Netherlands. *Limosa*, **39**, 182–6.

Étchécopar, R. D. & Hüe, F. (1978). *Les Oiseaux de Chine*. Les Editions du Pacifique, Tahiti.

Evans, P. R., Herdson, D. M., Knights, P. J. & Pienkowski, M. W. (1979). Short-term effects of reclamation of part of Seal Sands, Teesmouth, England, UK, on wintering waders and shelduck. 1. Shore bird diets, invertebrate densities and the impact of predation on the invertebrates. *Oecologia*, **41**, 183–206.

Evans, P. R. & Pienkowski, M. W. (1982). Behaviour of shelducks *Tadorna tadorna* (L.) in a winter flock: does regulation occur? *Journal of Animal Ecology*, **51**, 241–62.

Fielder, D. R. (1965). A dominance order for shelter in the Spiny Lobster *Lasus ialandei* (H. Milne-Edwards). *Behaviour*, **24**, 236–45.

Gallop, J. B. & Marshall, W. H. (1954). A guide to aging duck broods in the field. *Mississippi Flyway Council Technical Section Report*, 1954.

Gill, F. B. & Wolf, L. L. (1975). Economics of feeding territoriality in the Golden-winged Sunbird. *Ecology*, **56**, 333–45.

Gillham, E. H. (1950). Sheld-duck nesting amongst ground vegetation on saltings. *British Birds*, **43**, 190.

Gillham, E. H. (1951a). Down-stripping by Sheld-duck away from the nest site. *British Birds*, **44**, 103–4.

Gillham, E. H. (1951b). Voice of the Sheld-duck. *British Birds*, **44**, 103.

Gillham, E. H. & Homes, R. C. (1950). *Birds of the North Kent Marshes.* Collins, London.

Goethe, F. (1957). Uber den Mauserzug der Brandenten (*Tadorna tadorna* L.) zum Grossen Knechtsand. *Funfzig Jahre Seevogelschutz*, 96–106.

Goethe, F. (1961*a*). A survey of moulting Shelduck on Knechtsand. *British Birds*, **54**, 106–15.

Goethe, F. (1961*b*). The moult gatherings and moult migrations of Shelduck in North-west Germany. *British Birds*, **54**, 145–61.

Gorman, M. L. & Milne, H. (1972). Creche behaviour in the Common Eider *Somateria m. mollissima* L. *Ornis Scandinavica*, **3**, 21–5.

Goss-Custard, J. D. (1969). The winter feeding ecology of the Redshank *Tringa totanus*. *Ibis*, **111**, 338–56.

Gottlieb, G. (1965). Components of recognition in ducklings. *Natural History*, **74**, 12–19.

Grant, P. R. (1970). Experimental studies of competitive interaction in a two-species system: II, the behaviour of *Microtus, Peromyscus* and *Clethrionomys* species. *Animal Behaviour*, **18**, 411–26.

Greenhalgh, M. E. (1965). Shelduck numbers on the Ribble estuary. *Bird Study*, **12**, 255–6.

Grice, D. & Rogers, J. P. (1965). *The Wood Duck in Massachussetts.* Massachussetts Division of Fisheries and Game.

Hafner, H., Johnson, A. & Walmsley, J. (1979). The Camargue, France, Bird Report 1976–77. *La Terre et la Vie*, **33**, 307–24.

Harrison, J. (1967). *A Wealth of Wildfowl.* Deutsch, London.

Harrison, J. & Hudson, M. (1964). Some effects of severe weather on wildfowl in Kent in 1962–63. *Wildfowl Trust Annual Report*, **15**, 26–32.

Harrison, J. G. (1957). Avian tuberculosis in a wild Shelduck in association with an exceptional parasitic burden. *Bulletin of the British Ornithologists Club*, **77**, 149–50.

Harrison, J. M. (1953). *The Birds of Kent.* Witherby, London.

Harrison, J. M. (1965). Unilateral feather fragility in a wild Shelduck. *Bulletin of the British Ornithologists Club*, **86**, 36–9.

Hayne, R. A. (1949). Calculation of size of home range. *Journal of Mammalogy*, **30**, 1–18.

Hepburn, T. (1908). The Sheld-Duck in North Kent. *The Countryside*, Feb. 15, 1908.

Hilden, O. (1965). Habitat selection in birds. A review. *Annales Zoologici Fennici*, **2**, 53–75.

Hilden, O. & Vicolante, S. (1972). Breeding biology of the Red-necked Phalarope (*Phalaropus lobatus*) in Finland. *Ornis Fennica*, **49**, 57–85.

Hinde, R. (1970). *Animal Behaviour: A Synthesis of Ethology and Comparative Psychology.* McGraw-Hill, New York.

Hochbaum, H. A. (1959). *The Canvasback on a Prairie Marsh.* The Wildlife Management Institute, Washington.

Hoogerheide, C. & Hoogerheide, J. (1958). Slaggenrui van de Bergeend, *Tadorna tadorna* L., in Artis. *Ardea*, **46**, 149–58.

Hoogerheide, J. & Kraak, W. K. (1942). Voorkomen on trek von de Bergeend,

Tadorna tadorna L., naar aanleiding van veldobservaties aan de Gooije Kust. *Ardea*, **31**, 1–19.

Hori, J. (1962). Shelduck moulting in breeding area. *British Birds*, **55**, 418–9.

Hori, J. (1963). Observations on nesting Shelduck. *Wildfowl Trust Annual Report*, **14**, 168–9.

Hori, J. (1964a). The breeding biology of the Shelduck *Tadorna tadorna*. *Ibis*, **106**, 333–60.

Hori, J. (1964b). An automatic incubation recorder for wildfowl. *Wildfowl Trust Annual Report*, **15**, 97–9.

Hori, J. (1964c). Parental care in the Shelduck. *Wildfowl Trust Annual Report*, **15**, 100–3.

Hori, J. (1964d). Shelduck food supply in severe weather. *Wildfowl Trust Annual Report*, **15**, 44.

Hori, J. (1965a). Methods of distinguishing first year and adult Shelducks in the field. *British Birds*, **58**, 14–15.

Hori, J. (1965b). The display flights of Shelduck. *Wildfowl Trust Annual Report*, 1963–64, 58.

Hori, J. (1966). Moult migration of second-summer Shelduck. *Bird Study*, **13**, 99–100.

Hori, J. (1969). Social and population studies in the Shelduck. *Wildfowl*, **20**, 5–22.

Hüe, F. & Étchécopar, R. D. (1970). *Les Oiseaux du Proche et du Moyen Orient*. Editions N. Boubee & Cie., Paris.

Huxley, J. S. (1951). Communal display in the Sheld-duck. *British Birds*, **44**, 102–3.

Isakov, Y. A. (1955). *The Birds of the Soviet Union*. Volume 4. eds. G. P. Dementiev & W. A. Gladkov. In Special Review (D. D. Harber), *British Birds*, **48**, 404–10.

Jenkins, D. (1972). The status of Shelducks in the Forth Area. *Scottish Birds*, **7**, 183–201.

Jenkins, D., Murray, M. G. & Hall, P. (1975). Structure and regulation of a shelduck (*Tadorna tadorna* (L.)) population. *Journal of Animal Ecology*, **44**, 201–31.

Johnsgard, P. A. (1961). The taxonomy of the Anatidae – a behavioural analysis. *Ibis*, **103**, 71–85.

Johnsgard, P. A. (1965). *Handbook of Waterfowl Behaviour*. Cornell University Press, Ithaca.

Johnsgard, P. A. (1967). Dawn rendezvous on the lek. *Natural History*, **76**, 16–21.

Johnsgard, P. A. (1978). *Ducks, Geese, and Swans of the World*. University of Nebraska Press, Lincoln & London.

Kear, J. (1965). The internal food reserves of hatching Mallard ducklings. *Journal of Wildlife Management*, **29**, 523–8.

Kennedy, P. G., Ruttledge, R. F. & Scroope, C. F. (1954). *The Birds of Ireland*. Oliver & Boyd, Edinburgh & London.

King, B. (1960). Diving behaviour in Shelducks. *Wildfowl Trust Annual Report*, **11**, 156.

King, B. & Poulding, R. H. (1956). Shelduck breeding inland in Somerset. *British Birds*, **49**, 280.

King, J. (1973). Energetics of reproduction in birds. In *The Breeding Biology of Birds*, ed. D. S. Farner. National Academy of Sciences, Washington.

Kirkman, F. B. (1913). *The British Bird Book*. Volume 4. Jack, London.

Koskimies, J. & Lahti, L. (1964). Cold-hardiness of the newly hatched young in relation to ecology and distribution in ten species of European ducks. *Auk*, **81**, 281–307.

Knight, C. W. R. (1925). *Aristocrats of the Air*. Williams & Norgate, London.

Krebs, J. R. (1978). Optimal foraging: decision rules for predators. In *Behavioural Ecology*, eds. J. R. Krebs & N. B. Davies. Blackwell, Oxford.

Lack, D. (1964). A long-term study of the great tit (*Parus major*). *Journal of Animal Ecology*, **33**, 159–73.

Lack, D. (1974). *Evolution Illustrated by Waterfowl*. Blackwell, Oxford.

Ladhams, D. E. (1971). Behaviour of Shelduck in restricted territory. *Journal of the Bristol Ornithologists Club*, **1**, 176–7.

Lebret, T. (1976). Attacks by Avocets *Recurvirostra avosetta* on Shelduck *Tadorna tadorna*. *Limosa*, **49**, 17–23.

Leutz, C. P. & Hart, J. S. (1960). The effect of wind and moisture on heat loss through the fur of newborn Caribou. *Canadian Journal of Zoology*, **38**, 679–88.

Leyhausen, P. (1956). Verhaltensstudien an Katzen. *Zeitschrift fur Tierpsychologie*, **Supplement 2**.

Leyhausen, P. (1971). Dominance and territoriality as complemented in mammalian social structure. In *Behaviour and Environment; the Use of Space by Animals and Men*, ed. A. H. Esser, Plenum Press, New York.

Lind, H. (1957). A study of the movements of the Sheld-duck, *Tadorna tadorna* (L.). *Dansk Ornithologisk Forenings Tidsskrift*, **51**, 85–114.

Lind, H. & Poulsen, H. (1964). On the morphology and behaviour of a hybrid between Goosander and Shelduck (*Mergus merganser* L. × *Tadorna tadorna* L.). *Zeitschrift fur Tierpsychologie*, **20**, 558–96.

Lönneberg, E. (1932). Birds as 'relicts' in central Asia. *Ibis*, **2**, 625–32.

McAloney, K. (1973). *Brood ecology of the Common Eider (Somateria mollissima dresseri) in the Liscombe area of Nova Scotia*. Unpublished MSc dissertation, Acadia University, Nova Scotia.

McAtee, W. L. (1944). Sheld-duck in North Carolina. *Auk*, **61**, 148–9.

McDonald, D. (1980). Shelduck killed in territorial dispute. *Scottish Birds*, **11(2)**, 53.

Macdonald-Tyler, H. (1956). Shelduck, *Tadorna tadorna*, in Magilligan sandhills. *Irish Naturalists Journal*, **12**, 108.

MacFarland, C. (1972). Goliaths of the Galapagos. *National Geographic*, **142**, 633–49.

Maebe, J. & Vloet, H. van der (1952). Over rui, trek en biologie der Bergeend, *Tadorna tadorna* (L.) aan de Beneden-Schelde. *Gerfaut*, **42**, 59–83.

Makepeace, M. (1973). *Weather conditions and the survival of Shelduck ducklings on the Ythan estuary*. Unpublished MSc dissertation, Aberdeen University.

Makepeace, M. & Patterson, I. J. (1980). Duckling mortality in the Shelduck, in relation to density, aggressive interaction and weather. *Wildfowl*, **31**, 57–72.

Makkink, G. F. (1931). Die kopulation der Brandente (*Tadorna tadorna* (L.)). *Ardea*, **20**, 18–21.

Maksimoua, A. P. (1976). A new cestode *Fimbriarioides tadornae*, new species from the Sheldrake *Tadorna tadorna* and its development in the intermediate host. *Parazitologiya (Leningrad)*, **10**, 17–24.

Meiklejohn, M. F. M. (1950). Voice of sheld-duck. *British Birds*, **43**, 90.

Mendenhall, V. M. (1975). *Growth and mortality factors of Eider ducklings (Somateria m. mollissima) in north-east Scotland.* Unpublished PhD Thesis, Aberdeen University.

Mendenhall, V. M. (1979). Brooding of young ducklings by female Eiders *Somateria mollissima*. *Ornis Scandinavica*, **10**, 94–9.

Mertens, J. A. L. (1969). The influence of brood size on the energy metabolism and water loss of nestling Great Tits *Parus major major*. *Ibis*, **111**, 11–16.

Mertens, J. A. L. (1980). The energy requirements for incubation in Great Tits and other bird species. *Ardea*, **68**, 185–192.

Morley, J. V. (1966). The moult migration of Shelducks from Bridgwater Bay. *British Birds*, **59**, 141–7.

Morse, A. P. (1921). A Sheld duck (*Tadorna tadorna* L.) from Essex County, Mass. Franklin's Gull in New England. *Bulletin of the Essex County Ornithologists Club of Salem*, **1921**, 68–9.

Mosby, H. (1963). *Wildlife Investigational Techniques.* The Wildlife Society, Virginia.

Moss, R. (1969). A comparison of red grouse (*Lagopus l. scoticus*) stocks with the production and nutrition value of heather (*Calluna vulgaris*). *Journal of Animal Ecology*, **38**, 103–22.

Moss, R. & Watson, A. (1980). Inherent changes in the aggressive behaviour of a fluctuating Red Grouse *Lagopus lagopus scoticus* population. *Ardea*, **68**, 113–19.

Murton, R. K. (1966). A statistical evaluation of the effect of Woodpigeon shooting as evidenced by the recoveries of ringed birds. *The Statistician*, **16**, 183–202.

Murton, R. K. (1968). Some predator-prey relationships in bird damage and population control. In *The Problems of Birds as Pests*, eds. R. K. Murton & E. N. Wright. Academic Press, London & New York.

Murton, R. K., Isaacson, A. J. & Westwood, N. J. (1966). The relationships between woodpigeons and their clover food supply and the mechanism of population control. *Journal of Applied Ecology*, **3**, 55–96.

Nelson, D. (1944). Travelling Shelduck. *Country-side, Kingston-on-Thames Naturalists Society*, **12**, 272.

Newell, R. (1962). Behavioural aspects of the ecology of *Peringia* (*Hydrobia*) *ulvae* (Pennant) (Gasteropoda, Prosobranchia). *Proceedings of the Zoological Society of London*, **138**, 49–75.

Nice, M. M. (1941). The role of territory in bird life. *American Midland Naturalist*, **26**, 441–87.

Nisbet, I. C. T. & Vine, A. E. (1965). Regular inland breeding of Shelducks in the fens. *British Birds*, **48**, 362–3.

Noble, G. K. (1939). The role of dominance in the social life of birds. *Auk*, **56**, 263–73.

Odum, E. P. & Kuenzler, E. J. (1955). Measurement of territory size and home range size in birds. *Auk*, **72**, 128–37.

Oelke, H. (1969a). Die Brandgans (*Tadorna tadorna*) in Maussergebiet Grosserknechtsand. *Journal für Ornithologie*, **110**, 170–5.

Oelke, H. (1969b). Körpergewichte von Brandgänsen (*Tadorna tadorna*) im Mausergebiet Grosser Knechstand (Elbe-Wesermündung). *Vogelkundkliche Berichte aus Niedersachsen*, **1**, 47–50.

Oelke, H. (1970). Fresspuren von Brandgänsen im Mausergebiet Grosser Knechtsand (Elbe-Wessermündung). *Vogelwelt*, **91**, 107–11.

Oelke, H. (1974). Radiotelemetrische Untersuchungen an Brandgänsen (*Tadorna tadorna*) im Mausergebiet Grosser Knechtsand (Sommer 1973). *Journal für Ornithologie*, **115**, 181–91.

Olney, P. J. S. (1965). The food and feeding habits of Shelduck *Tadorna tadorna*. *Ibis*, **107**, 527–32.

Osieck, E. R. & Roselaar, C. S. (1972). A find of Marsh Sandpiper *Tringa stagnatilis* in the Netherlands. *Limosa*, **45**, 135–8.

Parslow, J. F. L. (1967). Changes in status among breeding birds in Britain and Ireland. *British Birds*, **60**, 2–47.

Parslow, J. (1973). *Breeding Birds of Britain and Ireland*. Poyser, Berkhamsted.

Patterson, I. J. (1976). The role of social behaviour in limiting the size of wildfowl populations and their output of young. *Wildfowl*, **27**, 140–1.

Patterson, I. J. (1977). Aggression and dominance in winter flocks of Shelduck *Tadorna tadorna* (L.). *Animal Behaviour*, **25**, 447–59.

Patterson, I. J. (1979). Tags and other distant-recognition markers for birds. In *Animal Marking. Recognition Marking of Animals for Research*, ed. B. Stonehouse, 54–62. Macmillan, London.

Patterson, I. J. (1980). Territorial behaviour and the limitation of population density. *Ardea*, **68**, 53–62.

Patterson, I. J., Gilboa, A. & Tozer, D. J. (in press a). Rearing other peoples' young: brood-mixing in the Shelduck *Tadorna tadorna*. *Animal Behaviour*.

Patterson, I. J. & Makepeace, M. (1979). Mutual interference during nest-prospecting in the Shelduck. *Animal Behaviour*, **27**, 522–35.

Patterson, I. J., Makepeace, M. & Williams, Murray (in press b). Limitation of local population size in the Shelduck. *Ardea*.

Patterson, I. J., Young, C. M. & Tompa, F. S. (1974). The Shelduck population of the Ythan estuary, Aberdeenshire. *Wildfowl*, **25**, 161–73.

Payn, W. H. (1962). *The Birds of Suffolk*. Barrie & Rockliff, London.

Perrett, D. H. (1951). Observations on Shelduck. *Report of the Mid-Somerset Naturalists Society*, **1**, 21–2.

Perrett, D. H. (1953). Shelduck observations 1952. *Report of the Mid-Somerset Naturalists Society*, **2**, 16–17.

Pienkowski, M. W. & Evans, P. R. (1979). The origins of Shelducks moulting on the Forth. *Bird Study*, **26**, 195–6.

Pienkowski, M. W. & Evans, P. R. (in press a). The disadvantages of colonial nesting for the Shelduck *Tadorna tadorna*. In *Proceedings of the International*

Waterfowl Research Bureau Symposium on Conservation of Colonially Nesting Waterfowl. Carthage, Tunisia, November 1978.

Pienkowski, M. W. & Evans, P. R. (in press *b*). Breeding behaviour, productivity and survival of colonial and non-colonial Shelducks *Tadorna tadorna* (L.) *Ornis Scandinavica*.

Pilcher, R. E. M. (1964). Effects of the cold winter of 1962–63 on birds of the north coast of the Wash. *Wildfowl*, **15**, 23–6.

Pitelka, F. A. (1959). Numbers, breeding schedule and territoriality in Pectoral Sandpipers in northern Alaska. *Condor*, **61**, 233–364.

Poulsen, H. (1957). Notes on the mating behaviour of the Sheld-duck (*Tadorna tadorna* (L.)). *Dansk Ornithologisk Forenings Tidsskrift*, **51**, 115–8.

Pulliam, H. R. (1980). Do Chipping Sparrows forage optimally? *Ardea*, **68**, 75–82.

Redhead, S. (1979). *Heat loss and brooding behaviour in relation to weather conditions in Shelduck ducklings*. Unpublished MSc dissertation, Aberdeen University.

Report on Somerset Birds. Shelduck *Tadorna tadorna* (L.). **1926**, 10; **1935**, 20; **1936**, 19–20; **1938**, 17; **1949**, 15–16.

Rittinghaus, H. (1956). Uber das verlassen der Brutohle und den Folgetreib bei jungen Brandenten. *Natur und Volk*, **86**, 168–73.

Roseveare, W. L. (1951). Quarrel between two families of Sheld-duck. *British Birds*, **44**, 390.

Salomonsen, F. (1968). The moult migration. *Wildfowl*, **19**, 5–24.

Saunders, H. (1889). *An Illustrated Manual of British Birds*. Gurney & Jackson, London.

Saunders, H., revised Clarke, W. Eagle. (1927). *Manual of British Birds*. Gurney & Jackson, London.

Schaller, G. B. (1967). *The Deer and the Tiger: A Study of Wildlife in India*. University of Chicago Press, Chicago.

Schein, M. W. & Fohrman, M. H. (1955). Social dominance relationships in a herd of dairy cattle. *British Journal of Animal Behaviour*, **3**, 45–55.

Schoener, T. W. (1968). Sizes of feeding territories among birds. *Ecology*, **49**, 123–41.

Seebohm, H. (1885). *A History of British Birds*. Volume 3. Porter, London.

Sharrock, J. T. R. (1976). *The Atlas of Breeding Birds in Britain and Ireland*. Poyser, Berkhamsted.

Shrub, M. (1979). *The Birds of Sussex. Their Present Status*. Phillimore, London & Chichester.

Siegel, S. (1956). *Nonparametric Statistics for the Behavioural Sciences*. McGraw-Hill, New York.

Slaney, P. A. & Northcote, T. G. (1972). Effects of prey abundance on density and territorial behaviour of young rainbow trout (*Salmo gairdneri*) in laboratory stream channels. *Journal of the Fisheries Board of Canada*, **31**, 1201–09.

Smith, C. (1844). *The Birds of Somersetshire*. Van Voorst, London.

South, S. R. & Butler, R. K. (1955). Shelduck breeding inland in Berkshire. *British Birds*, **48**, 277.

Southwick, C. (1953). A system of age classification for field studies of waterfowl broods. *Journal of Wildlife Management*, **17**, 1–8.

Sowls, L. K. (1955). *A study of the ecology and behaviour of some surface-feeding ducks*. Unpublished PhD thesis, University of Wisconsin.

Staton, J. (1945). Sheld-duck breeding in Nottinghamshire. *British Birds*, **38**, 356–7.

Stenger, J. (1958). Food habits and available food of Ovenbirds in relation to territory size. *Auk*, **75**, 335–46.

Stickel, L. F. (1954). A comparison of certain methods of measuring ranges of small mammals. *Journal of Mammalogy*, **35**, 1–15.

Suffern, C. (1951). Note of nestling Sheld-duck. *British Birds*, **44**, 391.

Swennen, C. & van der Baan, G. (1959). Tracking birds on tidal flats and beaches. *British Birds*, **52**, 15–18.

Taylor, J. (1976). The advantages of spacing-out. *Journal of Theoretical Biology*, **59**, 485–90.

Thompson, D. B. A. (1979). *Feeding behaviour of wintering shelduck on the Clyde estuary*. Unpublished BSc dissertation, Paisley College of Technology.

Thompson, D. B. A. (1981). A field estimation of prey intake by wintering shelduck (*Tadorna tadorna*). *Wildfowl*, **32**, 88–98.

Ticehurst, C. B. (1932). *A History of the Birds of Suffolk*. London.

Ticehurst, N. F. (1909). *A History of the Birds of Kent*. Witherby, London.

Tinbergen, N., Impekoven, M. & Frank, D. (1967). An experiment on spacing-out as a defence against predation. *Behaviour*, **28**, 307–21.

Tubbs, C. R. (1977). Wildfowl and waders in Langstone Harbour. *British Birds*, **70**, 177–99.

Walker, I. M. (1955). An earlier record of nesting Shelducks in Berkshire. *British Birds*, **48**, 277.

Walker, K. (1966). Mute swan taking over broods of Shelduck. *British Birds*, **59**, 432–3.

Walpole-Bond, J. A. (1938). *A History of Sussex Birds*. Witherby, London.

Walmsley, J. G. & Moser, M. E. (1981). The winter food and feeding habits of Shelduck (*Tadorna tadorna*) in the Camargue. *Wildfowl*, **32**, 99–106.

Watson, A. (1967). Social status and population regulation in the Red Grouse (*Lagopus lagopus scoticus*). *Proceedings of the Royal Society Population Study Group*, **2**, 22–30.

Watson, A. & Jenkins, D. (1968). Experiments on population control by territorial behaviour on red grouse. *Journal of Animal Ecology*, **37**, 595–614.

Watson, A. & Moss, R. (1972). A current model of population dynamics in red grouse. *Proceedings of the International Ornithological Congress*, **15**, 134–49.

Weller, M. W. (1959). Parasitic egg laying in the redhead (*Aythya americana*) and other North American *Anatidae*. *Ecological Monographs*, **29**, 333–365.

Williams, M. J. (1973). *Dispersionary behaviour and breeding of Shelduck Tadorna tadorna on the River Ythan estuary*. Unpublished PhD thesis, Aberdeen University.

Williams, M. J. (1974). Creching behaviour of the Shelduck *Tadorna tadorna* L. *Ornis Scandinavica*, **5**, 131–43.

Williams, M. J. (1975). Creching behaviour and survival of shelducklings. *Proceedings of the New Zealand Ecological Society*, **22**, 110–11.

Williams, T. S. (1953). Observations on Sheld-duck on the Welsh side of the Dee Estuary. *Proceedings of the Liverpool Naturalists and Field Club*, **92**, 15–18.

Williams, T. S. (1954). Further observations on the Sheld-duck on the Welsh side of the Dee Estuary. *Proceedings of the Liverpool Naturalists and Field Club*, **93**, 21.

Wilson, W. (1942). Field notes on the Common Sheld-duck. *North Western Naturalist, Arbroath*, **17**, 388.

Witherby, H. F., Jourdain, F. C. R., Ticehurst, N. F. & Tucker, B. W. (1939). *The Handbook of British Birds*. Volume 3. Witherby, London.

Wynne-Edwards, V. C. (1962). *Animal Dispersion in Relation to Social Behaviour*. Oliver & Boyd, Edinburgh & London.

Yarker, B. & Atkinson-Willes, G. L. (1971). The numerical distribution of some British breeding ducks. *Wildfowl*, **22**, 63–70.

Yarrell, W. (1843). *A History of British Birds*. Van Voorst, London.

Yarrell, W. & Saunders, H. (1884–85). *A History of British Birds*. Volume 4. Van Voorst, London.

Young, C. M. (1964a). *An ecological study of the Common Shelduck (Tadorna tadorna, L.) with special reference to the regulation of the Ythan population*. Unpublished PhD thesis, Aberdeen University.

Young, C. M. (1964b). Shelduck trapping methods. *Wildfowl Trust Annual Report*, **15**, 95–6.

Young, C. M. (1964c). Effects of recent hard winters on the Shelducks of the Ythan. *Wildfowl Trust Annual Report*, **15**, 45.

Young, C. M. (1970a). Territoriality in the Common Shelduck *Tadorna tadorna*. *Ibis*, **112**, 330–5.

Young, C. M. (1970b). Shelduck parliaments. *Ardea*, **58**, 125–30.

Subject index

272   *Subject index*

moult migration, 24–37
  flight on, 27–9
  arrival on, 29–33
  return from, 34–7
moulting areas, 20, 24–7
Mucula catena, 15
multiple clutches, 129, 142–5, 150–1, 155
mussels, 15
Mustela, 154
  erminea, 130
  nivalis, 155, 188
  vision, 242
mute swan, 175
Mya arenaria, 15
Mytilus edulis, 15

nearest-neighbour distances
  between broods, 173–4
  changes during winter in, 45–6
  in nesting area, 113
  of non-territorial males, 84
  between parents and young, 175–6
  of territorial males, 84
  in winter flock, 45–6
Nectarinia reichenowi, 92
Nereis diversicolor, 14–16, 47, 87, 93, 99, 172
nest-recorders, 147–8, 162
nest sites, 135–8
  in nest boxes, 137
  re-use of, 138
nesting success, 127–9, 150–6
  causes of failure in, 153–4
  in relation to density, 155–6
  timing of failure in, 152–3
  see also hatching success
nets
  cannon, 57
  clap, 57
non-breeding, 157–60

Oryctolagus cuniculus, 136

parasites, 243
parasitic laying, see multiple clutches
parental behaviour, 161–5, 182–5
'parliaments', 115–19
Parus major, 1
peregrine, 242
Phalaropus lobatus, 142
phenylenediamine, 59
Pica pica, 147
picric acid, 59
plumage, 6–7
population size
  of British population, 21–3
  in relation to breeding, 250–2
  of European population, 20–1

limitation of, 237–8, 252–5
local changes in, 238–41
in relation to survival, 250
predators
  of ducklings, 169, 188–9
  of nests, 138, 154–5
prospecting for nest sites, 107–37
  behaviour, during, 119–22
  costs and benefits of, 129–35
  interference between pairs, during, 124–9
  seasonal variations in, 109–11
  status of participants in, 110–12
  timing of, 107–9
Pteridium aquilinum, 136

rabbit, 136, 147, 155
range
  of broods, 170–1
  changes with season, 91
  overlap between, 93–5
  in nesting area, 116–19
  size, 88–92
recruitment, 218–36, 255
  of juveniles, 219–20
  limitation of, 230–2
  to the non-territorial flock, 228–32
  to territories, 232–4
red grouse, 3, 103
red-breasted merganser, 204, 209
red-necked phalarope, 142
redhead, 129
redirected attack, 183
removal experiments, 83, 103–5
Rissa tridactyla, 13, 154
rook, 115
Rubus fruticosus, 136
Ruppia, 15

Severn, 4, 20, 25, 27–8
sex ratio
  in the nesting area, 110–11
  in the non-territorial flock, 221–2
  in the winter flock, 43
shelduck
  description, 6–7, 212
  dimensions, 6, 20
  names, 6
  species, 5
  voice, 7–8, 119, 145, 161–2, 179
Sheppey, 4, 36, 44, 47, 87, 135, 138, 140, 143–6, 150–1, 155, 160, 162–5, 171, 179–82, 184, 188, 220, 240–1, 243, 250–3
site tenacity
  to brood ranges, 171
  to nesting areas, 117–18
  to territories, 83

Author index

Some key papers are referred to so frequently throughout the text that individual page references to their authors would not be very useful. These authors are: Buxton, N. E., Hall, P. (with Jenkins, D.), Hori, J., Jenkins, D., Makepeace, M. (with Patterson, I. J.), Murray, M. G. (with Jenkins, D.), Patterson, I. J., Williams, M. J. and Young, C. M.